OXFORD MEDICAL PUBLICATIONS

Spinal Cord Dysfunction: Assessment

Spinal Cord Dysfunction: Assessment

Edited by

L. S. Illis

Consultant Neurologist, Wessex Neurological Centre,
Southampton General Hospital,
Clinical Senior Lecturer in Neurology,
University of Southampton Medical School

Oxford   New York   Tokyo
OXFORD UNIVERSITY PRESS
1988

Oxford University Press, Walton Street, Oxford OX2 6DP
Oxford New York Toronto
Delhi Bombay Calcutta Madras Karachi
Petaling Jaya Singapore Hong Kong Tokyo
Nairobi Dar es Salaam Cape Town
Melbourne Auckland
and associated companies in
Berlin Ibadan

Oxford is a trade mark of Oxford University Press

Published in the United States
by Oxford University Press, New York

British Library Cataloguing in Publication Data
Spinal cord dysfunction: assessment
1. Spinal cord—Wounds and injuries
I. Illis, L. S.
617'.482044   RD594.3
ISBN 0 19 261624 2

Library of Congress Cataloging in Publication Data
Spinal cord dysfunction.
Includes bibliographical references and index.
1. Spinal cord—Wounds and injuries—Diagnosis.
I. Illis, L. S. (Leon S.)
RD594.3.S665   1988   617'.482075   87–31490
ISBN 0 19 261624 2

Set by Wyvern Typesetting Ltd, Bristol
Printed in Great Britain by
Butler & Tanner Ltd
Frome, Somerset

Foreword

Patrick D. Wall

The authors, editor, and the International Spinal Research Trust are to be congratulated on their initiative in producing this book. It is new and not even an update of old books. There are a number of interesting reasons why this book is new and needed.

It is hard to remember that, up until the Second World War, spinal cord transection was a terminal, lethal condition with the doctor's duty being to supervise the inevitable eventual death. Two factors combined to change that picture. One was the astonishing impact of the thinking, caring practice of Sir Ludwig Guttman, who was armed with the second new factor, antibiotics. The patients were dying of infections, and the neurological picture was confused since the effects of the cord injury itself and of the concomitant infections produced a mixed, shifting, deteriorating syndrome. It was incorrectly believed that cord transection removed essential trophic factors from paralysed tissue which condemned these organs to deteriorate progressively. Guttman's achievement was to deny this condemnation. He prevented deterioration through good nursing practice, by the use of sterile techniques and of antibiotics, by movement, and, above all, by raising the morale both of staff and of patients.

That success presented the medical staff, the scientists, and the patients with the problem of what to do next. That is the subject of this book. For the doctors it was possible for the first time to study the consequences of cord injury itself. We should particularly note the pioneering work of Prof. M. Dimitrijevic in this field. The old days of considering the labelling of the level of cord damage as an adequate diagnosis are gone. Few cord lesions are complete transections. Concomitant cord lesions below the level of major damage need diagnosis and treatment. The edge zone needs particular attention. Long-term progressive changes take place in the partially isolated cord. These changes are not necessarily inevitable but can be manipulated by therapy. These studies are not just academic; they define the problem exactly and set the stage, thus forming the basis for immediate therapy and for research.

The scientist has a crucial role beyond helping with techniques for the diagnostic–therapeutic assessments, important as they are. First he must carry out a detailed conversation with doctors and patients to understand exactly what the problem is. Too much basic science is directed at imagined clinical problems which do not relate to actual problems. Even with the enormously exciting prospects of tissue culture, transportation, and the induction of regeneration, the scientist must prepare to apply his discoveries to tissue as it exists in damaged cord. There is crucial work to be done short of the complete restoration of connection. For example, we do not know nearly enough about the progressive changes which take place in cells which have lost their input. There are certainly anatomical and functional changes in the cell and in near by intact structures which progress after nerve fibres have been cut. Quite a lot of work concentrates quite rightly on denervated peripheral structures, but almost none on central cells. Similarly we know little about the long-term consequences of changing the bombardment of central cells. Modern methods may well be capable of unravelling the sequence of changes, and perhaps their causes, and therefore their therapeutic manipulation.

Finally, and foremost, there is the patient, who is no longer a passive object to be probed and managed and treated. The patient needs to be recruited to be a member of his own treatment team. He and all those physicians and friends round him may hope and pray for a Lazarus-like resurrection, but in the meantime there is work to be done. The patient also wants to know exactly what it is that is wrong with him. He too needs to rank-order his problems in terms of which ones are the worst and which ones he wishes priority in overcoming, and then to match those priorities with what is feasibly correctable now or in the foreseeable future. The patient too must study and work to set the targets. This book is a beginning.

Preface

This book is based on a workshop held by the International Spinal Research Trust. The avowed aim of the Trust is to raise money for and encourage research in work leading to a cure for spinal injury. It may be surprising therefore to find no work within the covers of this book which mentions or leads to such a conclusion. The reason, of course, is simply that no such work exists at present, though within the next few years regeneration within the central nervous system (CNS), mammalian CNS transplants, and genetic engineering will undoubtedly change the present, somewhat gloomy view.

Twenty years ago, scientists who asked questions about CNS regeneration risked their reputations. Fortunately some scientists such as W. Windle in the USA and J. Z. Young in the UK never stopped asking such questions. The emphasis moved from the relatively fruitless search for regeneration as found in the peripheral nervous system to the way in which the intact nervous system reacts to a partial injury. The phenomena of sprouting of new synapses, unmasking of existing synapses, alteration of receptive fields, etc., has altered our view of the way in which the intact CNS functions, and raises questions pertinent to recovery and to the true nature of the effect of CNS damage in terms of disability. More recently, growth of damaged nerves within the CNS has been convincingly demonstrated and raises hopes even amongst the sceptical that 'cure' is perhaps not too over-optimistic, albeit in the somewhat distant future.

These exciting advances have scarcely filtered through to the clinical field, and have not been considered at all in terms of assessment of damage and disability. Yet accurate assessment must be the most essential prerequisite of neurological investigation both in clinical and research terms.

Dr Hans Frankel (pers. comm.) has emphasized that since potential therapies may be administered in the first few hours after injury, it is important to establish a body of information concerning prognosis in relation to an accurate neurological and neurophysiological examination in the early hours. At present most published work consists of statements such as 'within the first 24 hours' or 'within the first 48 hours'. We vitally need information concerning the first 3–6 hours. Because of the normal flow of patients through accident departments there is usually some delay before an accurate examination is performed. It is the responsibility of spinal injury units to acquire this information and to devise a workable and uniform classification of severity. The accumulated information will be invaluable in assessing future therapies. The classification and assessment must be based, however, on pathophysiology. It is for this reason that the workshop was convened and it is for this reason that this book has been published.

It should come as no surprise that plasticity features so large in this book since the theme is to draw the experimental evidence of pathophysiology and experimental physiology into the realm of practical clinical assessment. Plasticity has been variously summarized in functional or structural terms; e.g. as adaptive mechanisms by which the nervous system restores itself towards normal levels of functioning after injury (see Chapter 5) or the alteration of the system in such a way as to partially neutralize the change in condition (see Chapter 4). Similar descriptions could be adapted to describe rehabilitation; that is, rehabilitation is the means of restoring the patient towards normal levels of functioning and thus neutralizes the clinical effect of a lesion. The trouble with rehabilitation as a speciality, however, is that it rarely bases its ideas on neuroscience. Realistic assessment cannot be carried out without some understanding of pathophysiology, and intervention can only be rationally based on experimental neurology. In this sense assessment and treatment and rehabilitation become inseparable. It is no longer sufficient to provide only for the immediate medical needs of the patient and to treat complications.

This book represents the attempts of specialists in neurology, neurophysiology, spinal injury, neuropathology, and neuroradiology to indicate the difficulties and the possibilities of assessment and to emphasize the interdependence of clinical and experimental neurology.

In a book of this type there will be considerable overlap between chapters (by design) as well as omissions (both intentional and unintentional). I apologize to respected international experts not represented and offer the consolation that within the next few years the authors' work will, hopefully, be out of date in this rapidly expanding field of functional neurological and neurophysiological medicine.

Southampton
1987

L. S. Illis

Acknowledgements

This book is produced by and for the International Spinal Research Trust and it is a pleasure to acknowledge the unstinting help and advice of Mr Peter Banyard, the Research Director of the Trust.

My colleagues Dr Michael Sedgwick and Dr Jonathan Cole, of the Department of Clinical Neurophysiology at the Wessex Neurological Centre, have suffered interminable discussions patiently and have given invaluable ideas and time. I am deeply grateful to them both.

Finally, it is a pleasure to thank Mrs Jo Wilson for her many hours of patient secretarial work.

Contents

Contributors

J. D. Cole, Department of Clinical Neurophysiology, Wessex Neurological Centre, Southampton General Hospital, Southampton, UK.

P. Cook, Department of Neuroradiology, Wessex Neurological Centre, Southampton General Hospial, Southampton, UK.

M. R. Dimitrijevic, Department of Clinical Neurophysiology, the Institute for Rehabilitation and Research, Houston, and the Department of Rehabilitation, Baylor College of Medicine, Houston, Texas, USA.

M. G. Fehlings, Division of Neurosurgery and Playfair Neuroscience Unit, Toronto Western Hospital, University of Toronto, Toronto, Ontario, Canada.

J. T. Hughes, Radcliffe Infirmary, Oxford, and University of Oxford, Oxford, UK.

L. S. Illis, Wessex Neurological Centre, Southampton General Hospital, Southampton, and University of Southampton Medical School, Southampton, UK.

J. A. Massey, Department of Urology, Manchester Royal Infirmary, Manchester, UK.

P. Mayer, New York University Medical Centre, New York, NY, USA.

A. B. Rossier, Paraplegic Unit, University Orthopaedic Clinic, Balgrist, Zurich, Switzerland

J. C. Rothwell, Department of Neurology, Institute of Psychiatry, London, and King's College Hospital Medical School, London, UK.

E. M. Sedgwick, Department of Clinical Neurophysiology, Wessex Neurological Centre, Southampton General Hospital, Southampton, and University of Southampton Medical School, Southampton, UK.

M. Swash, Department of Neurology, The London Hospital, and St Mark's Hospital, London, UK.

C. H. Tator, Division of Neurosurgery and Playfair Neuroscience Unit, Toronto Western Hospital, University of Toronto, Toronto, Ontario, Canada.

P. D. Wall, Cerebral Functions Research Group, Department of Anatomy and Embryology, University College, London, UK.

C. J. Woolf, Cerebral Functions Research Group, Department of Anatomy and Embryology, University College, London, UK.

W. Young, Department of Neurosurgery, New York University Medical Centre, New York, NY, USA.

I

Structural and functional changes

A review of models of acute experimental spinal cord injury

Michael G. Fehlings and Charles H. Tator

Introduction: an historical perspective

The earliest descriptions of spinal cord injury in man were made by Egyptian physicians *c.*2500 BC, as recorded in *The Edwin Smith surgical papyrus* (Breasted 1930). Galen, who sectioned the spinal cord of primates and other animals, is reported to have conducted the first studies of experimental cord injury (Siegel 1973). Leonardo da Vinci performed experiments on frogs and concluded that within 'the spinal cord ... lies the foundation of movement of life' (Riese 1959). It was not until the late nineteenth century that researchers examined the pathological mechanisms involved in cord injuries. In 1890, Schmaus applied direct blows to the backs of rabbits and noted areas of degeneration and cavitation within the spinal cord. In 1899, Spiller studied a cat subjected to pressure crudely applied to the spine and reported degeneration of the anterior and anterolateral white matter of the cord.

The modern era of cord injury research was heralded by Allen in 1911, who introduced a model of experimental spinal cord injury which could be quantified and standardized. Allen dropped weights from varying heights onto the exposed cord of dogs, and expressed the force of these contusion injuries in gram centimetres (g cm). He reported that a 300 g cm injury caused 'temporary paraplegia' whereas a 400 g cm injury resulted in permanent paraplegia. Intramedullary haemorrhage and oedema were maximal four hours after the injuries were sustained. Allen's 'weight-dropping method' became one of the most frequently used cord injury models, and there have been numerous modifications of it, as described below, to overcome its major shortcomings. Since the pioneering work of Allen, several other models have been developed to simulate the numerous mechanisms of spinal cord injury in humans. Some have the necessary precision and consistency required for randomized, controlled, therapeutic trials, while others allow the isolation for study of specific aspects of cord injury pathophysiology.

Models of experimental spinal cord injury can be classified as follows: (1)

Table 1.1. Classification of experimental models of acute spinal cord injury

	General models	Specific models
Purpose	To study the pathophysiology and treatment of clinically relevant cord injuries	To study isolated factors of spinal cord injury pathophysiology and treatment

Types	A. *Compression models*
	    I. Kinetic compression
	    (a) Weight-drop
	    (b) Clip compression
	    (c) Rapid balloon inflation
	    (d) Vertebral dislocation
	    II. Static compression
	    (a) Slow-graded cord compression
	    (b) Slowly expanding extradural mass lesions
	 B. *Acceleration–deceleration*
	 C. *Transection*
	 D. *Vertebral distraction*
	 E. *Ischaemic injuries due to vascular occlusion*
		A. *In vivo*
		    I. Chemical injury
		    II. Models to study axonal regeneration
		 B. *In vitro*
		    I. Preparation of isolated spinal cord axons
		    II. Tissue culture

general models which simulate human cord injuries; and (2) specific models to study isolated factors of cord injury pathophysiology (Table 1.1). In the next section, the general injury models will be discussed, with particular reference to their clinical relevance and reproducibility of the injuries inflicted. Then the specific models and their specialized applications will be discussed and, finally, there will be a critical examination of the techniques used to evaluate the severity and outcome of experimental spinal cord injury.

General models of spinal cord injury

In most countries, the common causes of human cord injuries include motor vehicle accidents, sports/recreational injuries, work-related accidents, and falls at home (Tator and Edmonds 1979), while in some countries, including the USA, missile injuries due to fire-arms are also common (Kraus, Franti, Riggins, Richards, and Borhani 1975). In motor vehicle accidents and sports/recreational accidents, currently the most frequent causes of cord injuries in most countries, the usual underlying mechanism is acute cord compression due to bone displacement accompanying fracture–dislocation or burst fracture (Tator 1983). Acute stretching and acceleration–deceleration are other less common mechanisms. Although the spinal cord is rarely anatomically transected initially, the physical forces of severe trauma including compression, laceration, or stretching can initiate a series of secondary pathological changes that ultimately results in virtually complete transection (Tator and Rowed 1979). Several models which attempt to mimic these forces and pathological changes have been developed.

Compression models
Compression models of cord injury can be classified as kinetic or static according to the biomechanics of the applied forces. Kinetic compression models involve rapid compression of the cord in less than one second. Indeed, most kinetic models compress the cord in less than 100 milliseconds. Furthermore, the applied force compresses the cord with increasing velocity (acceleration greater than zero) to the point of maximal cord compression. In contrast, static compression models use forces which slowly compress the cord (i.e. over a period of more than one second) at approximately constant velocity (zero acceleration).

Kinetic compression models Essentially, four types of kinetic compression models have been developed: (1) the weight-drop technique, (2) the extradural balloon compression technique, (3) the clip compression method, and (4) the vertebral dislocation model.

The Freeman and Wright (1953) modification of Allen's (1911) weight-drop technique has been the most widely used method for producing experimental spinal cord injury (Fig. 1.1). A weight is allowed to fall a given distance through a vertically oriented tube to strike an impounder resting directly on the spinal cord exposed through a laminectomy. The resulting force is expressed in g cm, which is an arbitrary term, derived from the product of weight and height, which does not reflect the kinetic energy transferred to the spinal cord (Dohrmann, Panjabi, and Banks 1978). Dorhmann and Panjabi (1976) showed that the energy imparted by a 50 g mass falling 10 cm was 100-times greater than that of a 5 g mass falling 80 cm, although both represented '400 g cm' injuries.

The weight-drop method has a number of disadvantages. For example, it causes posterior cord compression, whereas in patients compression is usually due to circumferential cord compression from fracture–dislocation or to anterior compression from burst fractures (Tator 1983). More importantly, the method has been reported to produce considerable variability in clinical outcome (Koozekanani, Vise, Hashemi, and McGhee 1976; Khan and Griebel 1983) and in the pathology of the injury site (Dohrmann and Panjabi 1976; Khan, Griebel, Rozdilsky, and Politis 1985).

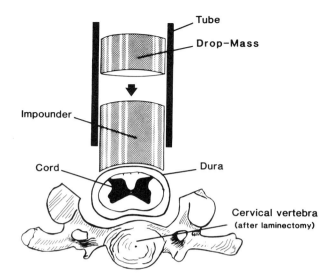

Fig. 1.1. The weight-drop model of cord injury.

A number of modifications to the weight-drop method have appeared in an attempt to correct this variability. Ford (1983) placed a curved plate anterior to the cord to provide a smooth, hard surface for the receipt of the posterior impact and, in addition, used a magnet to prevent the weight from repeatedly striking the cord surface during rebound as occurs with other weight-drop models. Other authors have stressed the need to stabilize the vertebral column to inflict more reproducible injuries (Blight 1985). One group (Bresnahan, Todd, Noyes, and Beattie 1985) recently described the use of a feedback-controlled magnetically driven impactor to standardize the injury force, and another developed a pneumatically driven piston, the tip of which rapidly compressed the dorsal surface of the cord and recoiled back at a speed of 2 m/s (Anderson 1982).

In 1957, Tarlov described the use of an extradural inflatable balloon to inflict spinal cord injuries in dogs. Through a laminectomy, the balloon was inserted into the spinal cord and fed into an adjacent area with intact lamina. Although the duration of compression could be accurately controlled, the balloon could not be positioned consistently, and therefore the pressure on the cord could not be accurately controlled. In an effort to improve this technique, Tator (1972) placed an extradural inflatable Silastic cuff around the spinal cord of monkeys exposed through a laminectomy. The cuff could be inflated to varying pressures within the cuff and left in place for varying periods of time. For example, rapid inflation of the cuff to a pressure of 350 mm Hg for five minutes resulted in complete post-operative paraplegia, although five of the ten monkeys achieved partial neurological recovery by twelve weeks. The pathology at the injury site, after twelve weeks, consisted of central cavitation, gliosis, and white matter demyelination. Hukuda, Mochizuki, and Ogata (1980) modified this technique by applying a circumferential, extradural tourniquet made from a Penrose drain and inflating it rapidly.

In the past, these balloon or cuff compression techniques have required the use of large animals, which are more expensive and less readily available than small animals such as the rat. However, Khan and Griebel (1983) have recently used a small Fogarty arterial embolectomy balloon catheter to produce cord injuries in rats. The spinal cords of rats subjected to 0.1 ml balloon compression for either three or five minutes showed no histological changes or only mild degenerative neuronal changes in the grey matter, whereas the spinal cords compressed for seven minutes with 0.1 ml compression showed extensive cavitation and necrosis at the injury site. The balloon and cuff techniques have the serious disadvantage that, although the pressure in the device can be measured, the pressure on the cord is extremely difficult to measure in large animals, and probably impossible to measure in small animals.

In our laboratory, Rivlin and Tator (1978*a*) described a new model of cord injury in rats which used a modified Kerr–Lougheed aneurysm clip to inflict an extradural compression injury on the cord (Fig. 1.2). With an aneurysm clip applicator, the blades of the clip were passed extradurally anteriorly and posteriorly around the exposed cord and then the clip was *rapidly* released from the applicator to produce immediate, acute cord compression. The major advantages of this method are that the force of closure of the clip can be calibrated precisely (Dolan and Tator 1979) and the strength of the clip's C-shaped steel spring can be varied to produce clips of various forces. Furthermore, the duration of compression of the cord can be easily and accurately varied to produce spinal cord injuries of differing severity (Dolan, Tator, and Endrenyi 1980). This model is marked by its simplicity, its usefulness in small animals, its reproducibility, and, since the force of compression is applied both anteriorly and posteriorly to the cord, its relevance to clinical cord injuries in man. In contrast, the variability of the cord injuries produced with the weight-drop technique is magnified when small animals such as the rat are used (Khan and Griebel 1983).

The weight-drop method, the extradural balloon compression technique, and the clip compression method have recently been compared in rats with respect both to the consistency of clinical outcome (Khan and Griebel 1983) and to the pathology at the injury site (Khan *et al.* 1985). The weight-drop technique resulted in considerable variability both in clinical outcome and in injury pathology and 'was found unreliable for experimental spinal cord injury in the rat' (Khan and Griebel 1983). The clip compression model

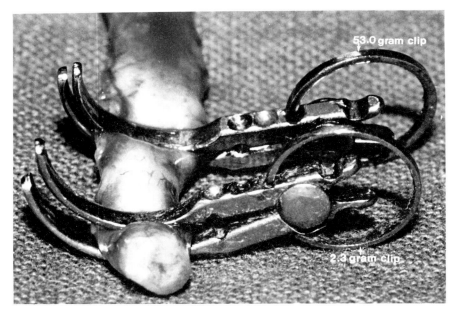

Fig. 1.2. The clip compression model of cord injury. Two clips exerting forces of 2.3 g and 53.0 g, respectively, are shown compressing an *ex situ* rat cervical cord. The blades of the stronger clip are compressing the cord to a greater degree.

resulted in the most consistent cord injuries with respect to clinical recovery and produced the least variability in pathological changes. As noted above, rats subjected to a 0.1 ml balloon compression injury for either three or five minutes showed a complete clinical recovery, while a seven-minute compression with the same force resulted in complete and irreversible paraplegia. This steep dose–response curve represents a major disadvantage of the balloon compression technique in rats, namely the difficulty in producing a submaximal injury. In contrast, the clip compression technique can generate very consistent and graded dose–response curves with respect to both duration and force of compression (Dolan *et al.* 1980).

A kinetic compression model of cord injury has been developed which does not require exposure of the cord through a laminectomy and hence accurately mimics human cord injuries. Fialho, Lumb, and Scott (1982) used a pneumatically powered device to inflict spinal cord injury in dogs by producing a traumatic subluxation between the L1 and L2 vertebrae. The displacement was accomplished in less than one second and required a force exceeding 2000 N. A serious limitation of the model, however, is its inability to accurately control the severity of the cord injury.

Static compression models   Static compression models use a slowly applied compressive force to injure the cord and attempt to simulate mass lesions which slowly compress the cord such as extradural tumours. Essentially two types of lesion-making methods can be considered in this category: (1) the technique of graded additions of compression and (2) the use of slowly expanding mass lesions.

In an early, crude attempt to induce a static compression injury of the cord, McVeigh (1923) applied finger pressure to the exposed spinal cord of dogs which resulted in haemorrhage into the grey matter and oedema of the lateral and anterior funiculi of the cord. In 1957, Tarlov described a method of gradual cord compression in dogs by slowly inflating an extradural balloon at the rate of 0.25 cm^3. The balloon was introduced via a laminectomy at T12 and threaded extradurally to the 'mid-thoracic region'. The animals became paraparetic within forty-five minutes and paraplegic within two hours. Unfortunately, as with Tarlov's method of kinetic cord compression, it was very difficult to accurately position the extradural balloon. Croft, Brodkey, and Nulsen (1972) applied weights directly onto the cord exposed through a laminectomy to reversibly block both motor and somatosensory-evoked potentials (SSEPs). Eidelberg, Staten, Watkins, McGraw, and McFadden (1976) applied weights sequentially to the thoracic cord of ferrets to produce a model of incomplete cord injury. Hukuda and Wilson (1972) and Schramm, Shigeno, and Brock (1983) have produced slow, graded compression of the cord by means of a screw-plate assembly with stepwise advancement of the compression device.

A variety of slowly expanding extradural mass lesions have been used to produce experimental cord injury. McCallum and Bennett (1976) inserted casein, a hydroscopic plastic which slowly absorbs water and can double its weight and volume, extradurally between the thoracic vertebrae and cord to produce cord compression sufficiently severe to result in loss of the SSEP. To simulate extradural tumours which cause cord compression, Ushio, Posner, Kim, Shapiro, and Posner (1977) placed Walker 256 carcinoma cell suspensions between the cord and the T12 vertebral body. The resultant tumour produced cord compression with paraplegia within four weeks.

Although the static compression models have the major advantage of being easy to use, and highly reproducible, they do not simulate the biomechanical features of most types of cord injury in patients and should not be used to model acute spinal cord injury. The static compression models do not seem to precipitate the vascular damage characteristic of the kinetic compression models, which is an inherent component of acute spinal cord injuries in patients.

Acceleration–deceleration cord injuries

Unterharnscheidt (1983) described a model of acceleration–deceleration spinal cord injury in primates. A conscious rhesus monkey was strapped to a sled and subjected to rapid acceleration and deceleration to simulate a high-speed automobile accident. At an acceleration–deceleration force of 110–120 G, half the animals sustained an atlanto-occipital separation or C1–2 subluxation, and the majority sustained a high cervical cord compression or transection. Although this model mimics the biomechanics of severe human cord injuries, it produces injuries of variable severity and location in the spinal cord.

Cord transection

Since patients with spinal cord injury rarely show anatomical transection of the cord initially (Tator and Rowed 1979), transection is not an appropriate model for studying the pathophysiology or treatment of acute spinal cord injury. Although transection models are useful in studying regeneration of the spinal cord, transection of the cord even with a sharp knife can produce considerable mechanical damage, resulting in spreading haemorrhage and necrosis. To avoid this problem, Barrett, Donnati, and Guth (1981) used a CO_2 laser to transect the rat spinal cord and reported less morphological damage than with a thin blade. The neodynium yttrium–aluminium–garnet laser and the argon laser both achieve better haemostatis than the CO_2 laser but are reported to cut tissue less precisely (Edwards, Boggan, and Fuller 1983).

Cord injury produced by vertebral distraction

Distraction is one of the less commonly recognized mechanisms of human cord injury because it is more difficult to document. Indeed, almost all the reported cases of distraction injury have been iatrogenic: in obstetrics due to excessive traction during difficult breech deliveries; in orthopaedics during traction or Harrington distraction for scoliosis; and in neurosurgery during cervical traction for spinal injury (Dolan, Transfeldt, Tator, Simmons, and Hughes 1980).

A number of researchers have developed models of spinal cord injury involving vertebral distraction. In our laboratory, we investigated the pathophysiology of traction injuries by applying a distraction apparatus to the L2 and L3 vertebral bodies of the cat (Dolan *et al.* 1980). We found that when the distraction was severe enough to alter the spinal-evoked potential at T13, there was already a 50 per cent decrease in spinal cord blood-flow (SCBF). We concluded that spinal cord ischaemia was primarily responsible for cord injury following vertebral distraction. Cusick, Myklebust, Zyvoloski, Sances, Houterman, and Larson (1982) applied skull traction to monkeys and recorded the SSEPs from multiple sites as well as the SCBF.

They found that 75–100 lb of gradually applied axial traction resulted in altered SSEPs, and it was postulated that these early changes were the result of stretching of the fibre tracts rather than alteration of SCBF. Another model of distraction injury of the cord was reported by Yamada, Knierim, Maeda, and Schulz (1981), who applied traction to the cat spinal cord by pulling on the *filum terminale* to simulate the tethered cord syndrome and noted redox changes within the cord indicative of anoxia, but did not measure SCBF in this experiment.

Ischaemic injury due to vascular occlusion

Bastian first described the syndrome of spontaneous spinal cord infarction in 1886, and Ross (1985) recently reviewed the subject of spinal cord infarction following surgery of the aorta. Operative repair of thoracic aortic aneurysms carries a 6 per cent risk of paraplegia, and spinal cord infarction has also been described following repair of aortic coarctation, resection of abdominal aortic aneurysms, and aortic dissection. Accordingly, this important clinical problem has given rise to several experimental models designed to simulate these clinical situations.

Spinal cord blood-flow is complex and extremely variable, both within a given species and between species. For example, dogs, guinea pigs, and, to a lesser extent, rats have extensive collaterals to the cord from the thoracic aorta (Woollam and Millen 1955) and, therefore, infrarenal ligation of the aorta does not produce paraplegia in these animals. However, cats and rabbits will show consistent spinal cord infarcts following infrarenal occlusion of the aorta. For example, Zivin and DeGirolami (1980) demonstrated that the caudal spinal cord of the rabbit receives its vascular supply primarily from a branch of the abdominal aorta. After an aneurysm clip was placed for sixty minutes on the aorta distal to the left renal artery there was complete and irreversible paraplegia accompanied by the pathological findings of central grey matter cavitation and necrosis of the dorsal horns. Krogh (1950) occluded the rabbit's abdominal aorta with an external clamp, but unfortunately this caused visceral damage and inconsistent degrees of spinal cord ischaemia.

Anderson, Behbehani, Means, Waters, and Green (1983) used a Fogarty balloon catheter to occlude the thoracic aorta of cats, and when the blood pressure cephalad to the balloon was reduced to 65 mm Hg by exsanguination, total ischaemia of the cord was induced. Thirty minutes of severe cord ischaemia significantly reduced the spinal cord concentrations of glucose and high-energy phosphates, but, if the cord was reperfused, the levels of these metabolites returned to normal.

Another approach has been to occlude the spinal cord blood vessels directly. For example, Doppman, Girton, and Popovsky (1979) passed a number-30 lymphangiography needle directly into the posterior spinal vein of adult monkeys and occluded the vessel with silicon rubber. Also, Fried, di Chiro, and Doppman (1969) produced selective occlusions of medullary arteries including the great medullary artery of Adamkiewicz.

One of the major shortcomings of most of the experiments based on these models of ischaemic spinal cord injury is the lack of concomitant measurement of SCBF. The work by Anderson *et al.* (1983) is a notable exception. Such measurements are essential both to establish that the cord has been rendered ischaemic by the model and to determine whether the therapy under question is of value in restoring or preserving SCBF. 'Occlusion' of the vascular supply without measurement of actual SCBF is a serious methodological error.

Specific models to study isolated factors of spinal cord injury pathophysiology

The general models of spinal cord injury described above simulate spinal cord injuries in humans and are useful in assessing the therapeutic value of experimental treatments and in investigating factors involved in the pathogenesis of cord injury. To study specific factors involved in the pathophysiology of spinal cord injury and treatment, specialized models have been developed which can be classified into those using *in vivo* or *in vitro* techniques (Table 1.1).

In vivo *models*
This group can be divided into models for studying the injurious effects of specific toxins (chemical injuries) and those for investigating axonal regeneration in the spinal cord.

Chemical injury models A number of models have been developed for studying the toxic effects on the spinal cord of epinephrine, norepinephrine, calcium chloride ($CaCl_2$), adrenergic neurotoxins, and serotoninergic neurotoxins. For example, a dose of 0.3–0.75 mg epinephrine injected into the lumbar subarachnoid space of the rabbit caused chromatolysis with fragmentation of the Nissl granules, shrinkage of dorsal and ventral horn cells, and demyelination of the dorsal columns (Berman and Murray 1972). Furthermore, Osterholm and Mathews (1972) found that $35 \mu g$ of norepinephrine injected directly into the central grey matter of the cat spinal cord resulted in haemorrhage and oedema within two hours, and suggested that altered norepinephrine metabolism may be a factor in the pathogenesis of cord injury.

Balentine and Dean (1982) developed a model of $CaCl_2$-induced myelopathy which has proven useful in studying the deleterious effects of calcium on the spinal cord. One millilitre of $CaCl_2$ (pH 7.4) dripped onto the spinal cord for ten minutes produced paraplegia, spongiosis of the spinal cord, and necrosis at the site of application. Banik, Hogan, Powers, and Whetstine (1982) used this model of $CaCl_2$-induced myelopathy to demonstrate progressive loss of neurofilament, microtubular, and glial filament proteins over a period of eight hours to five days following injury.

When injected directly into the cord, the neurotoxins 6-hydroxydopamine and 5,6-dihydroxytryptamine cause selective destruction of adrenergic and serotoninergic neurons and axons (Nobin, Baumgarten, Björkland, Lachenmayer, and Stenevi 1973), and thus this selective chemical axotomy avoids the need for transecting the cord in studies of axonal regeneration. A criticism of the technique, however, is that the damage may not be restricted to the specific neurotransmitter system, although a number of authors have used this technique effectively in studies of axonal regeneration (e.g. Nornes, Bjorklund, and Stenevi 1983).

Models used to study central axonal regeneration In 1980, the Advisory Task Force of the National Institute of Neurological and Communicative Diseases and Stroke published a series of criteria for experimental models used in the study of spinal cord regeneration:

1) the experimental lesion must cause disconnection of nerve processes; 2) processes of central nervous system neurons must bridge the level of injury; 3) the regenerated fibres must make junctional contacts; 4) the regenerated fibres must generate post-junctional responses; and 5) changes in function must derive from regenerated connections (Guth, Brewer, Collins, Goldberger, and Perl 1980).

Although it is doubtful that a given study can fulfil all the above criteria, one should strive to satisfy as many as possible in experimental models to be used in evaluating central axonal regeneration.

Many different models of cord injury and several different therapeutic strategies have been used to study axonal regeneration. They can be grouped into three main types:

(1) grafts or matrices to bridge the site of partial or complete spinal cord transection (Richardson, McGuiness, and Aguayo 1980, 1982; David and Aguayo 1981; Bunge, Johnson, and Thuline 1983; Richardson, Issa, and Aguayo 1984);
(2) shortening of the spinal column to appose the two cut stumps of the cord (Derlon, Camille-Roy, Lechevalier, Bisserie, and Coston 1983);
(3) the use of electrical stimulation (Borgens, Roederer, and Cohen 1981; Wallace, Tator, and Piper 1984) or growth factors, such as Schwann cell extracellular matrix (Madison, da Silva, Dikkes, Chui, and Sidman 1985), to promote regeneration of transected or traumatized axons.

In general, models which involve complete transection of the cord are favoured in studies of central axonal regeneration to avoid the pitfall of mistakenly identifying persisting fibres as regenerating fibres. Experiments involving 'subpial dissection' (Kao 1974) should be avoided because claims of regeneration cannot usually be substantiated. When incomplete lesions are made, great care must be taken to evaluate outcome, as described below.

In vitro *models*

Two *in vitro* models have been developed to study specific aspects of the pathophysiology of spinal cord injury: (1) preparations of isolated segments of spinal cord axons, and (2) tissue culture.

Preparations of isolated spinal cord axons   Thin slices of central nervous system tissue maintained in artificial media have provided standard preparations for many types of physiological study, although application of this technique to the spinal cord has been limited. Rudin and Eisenman (1951) showed that it was possible to isolate white matter tracts from adult cat spinal cord and to maintain them in an oxygenated bath of physiological saline for studies of axonal conduction. Blight (1983*a*) dissected out the dorsal columns from the thoracic cord of normal cats and of paraplegic cats with chronic spinal cord injuries and maintained them in oxygenated Krebs' solution. With microelectrodes he then recorded action potentials from single axons for up to fifteen hours. The *in vitro* conduction properties of axons from the normal cord were comparable to *in vivo* physiological data. Interestingly, 7 per cent of the axons sampled from the cords of chronic paraplegic cats were able to conduct action potentials, although the conduction properties were highly abnormal.

   This type of experiment has considerable scope for advancing our knowledge of the pathophysiology and treatment of central axonal injury. The results, however, would always require verification by one of the general models.

Tissue culture   The most distinct advantage of the tissue culture technique is that it allows precise control of the experimental milieu. However, due to the isolation of the tissue culture system, the results cannot be extrapolated directly to *in vivo* systems. Nevertheless, valuable information regarding the growth and behaviour of cultured neurites and supporting cells has aided spinal cord injury research. For example, the influence of direct current on the growth of axons has been elucidated using cultured neurites (Jackson, Lecar, Brenneman, Fitzgerald, and Nelson 1982; Patel and Poo 1982, 1984). Moreover, tissue culture is an excellent method for studying the influence of various trophic factors such as laminin and fibronectin on neurite growth (Ard, Bunge, and Bunge 1985; Coughlin, Grover, and Jung 1985).

Methods used to evaluate outcome after experimental spinal cord injury

Evaluation of outcome is an extremely important aspect of experimental models of acute spinal cord injury. To be useful for the elucidation of basic pathophysiological data or for the evaluation of therapeutic strategies, a model must allow the generation of accurate, easily obtained, and readily interpretable dose–response curves involving one or more outcome parameters. The outcome parameters should be objective, consistent, and quantifiable. The techniques for evaluating outcome after experimental spinal cord injury are summarized in Table 1.2.

Table 1.2 Evaluation of the outcome of experimental spinal cord injury

Technique	Types
Clinical neurological evaluation	Tarlov scale
	Modifications of Tarlov scale
Functional clinical tests	Grid-walking
	Inclined ramp
	Inclined plane
Tests of structural integrity	Histopathology
	Angiography
	Imaging
Spinal cord blood-flow measurements	
Axonal tracers	
Biochemical evaluation	
Neurophysiological tests	H-reflex
	Evoked potentials —spinal
	—somatosensory
	—vestibulospinal
	—motor

Clinical neurological evaluation

In patients the clinical neurological evaluation is the most important method of assessing the outcome of spinal cord injury, especially for determining the value of a proposed treatment. In 1957, Tarlov reported a grading scale of clinical neurological function in dogs based on examination following cord injury, and since then various modifications of Tarlov's scale have been devised. In general, they have been the most frequently used methods for assessing somatic motor function below the level of a cord lesion. Although Tarlov's original scale also evaluated sensory function and bladder function, most of the subsequent modifications have been limited to somatic motor function (Means, Anderson, Waters, and Kalaf 1981). In primates such a grading scale is reasonably accurate, since primates with cord transection are incapable of complex motor functions such as walking or stepping (Rivlin and Tator 1977). Indeed, our laboratory has had extensive experience with this method of evaluation in primates (Tator 1972). However, lower species of mammals, such as cats and dogs, with complete spinal cord transection are capable of 'spinal walking' and other forms of reflex limb movement which can mimic voluntary motion, and these interfere with the clinical neurological assessment of limb function (Eidelberg *et al.* 1976; Rivlin and Tator 1977). Indeed, Grillner (1973, 1975) has shown that cord-transected cats can support their body weight on their hindlimbs. In our opinion the Tarlov scale and its modifications should not be used in species other than primates.

Functional clinical tests

To overcome the above limitations, a number of investigators have proposed alternative methods for the clinical evaluation of the neurological status of experimental animals. Wilson (1984) assessed the ability of cord-

Fig. 1.3. A normal rat on the inclined plane maintaining its position at an angle of 75°.

injured cats to traverse a grid and recorded their performance on film. This technique, however, lacks objectivity and quantifiability. Eidelberg *et al.* (1976) used an inclined wooden ramp to assess cord-injured ferrets. Prior to cord injury, the animals were trained to climb the ramp for food rewards and the criterion adopted for adequate pre-operative performance was the ability to climb a 25° slope in 1.0 ± 0.2 seconds. After cord injury, there was a correlation between the percentage of white matter spared by the injury and the animal's performance. In our laboratory, we devised a similar technique for rats (Rivlin and Tator 1977) which avoided the need for reinforcement and operant conditioning (Fig. 1.3). After cord injury, the rats are placed horizontally on an inclined plane which can be adjusted to provide a slope of varying grade. The score assigned to the rat is the maximum angle of the plane at which the animal can maintain its position for five seconds without falling. Normal rats achieve scores of 80.5° ± 3.6°, while rats with cord transections at C7 maintain themselves at an angle of only 23.0° ± 2.9° (Rivlin and Tator 1977). This method has the advantages of being accurate, consistent, and quantifiable.

The above methods of clinical assessment primarily reflect motor function. Indeed, it is difficult with clinical testing to accurately assess sensory function in non-primate experimental animals because of the presence of reflex limb movements, which are mediated at a spinal cord level, in response to noxious stimuli. SSEPs (see below), however, can be used to assess dorsal column integrity in a quantifiable manner.

Structural integrity tests for the anatomical assessment of the site of cord injury

The extent of the pathological changes at the site of cord injury can be evaluated using a number of techniques including histopathology, angiography, and imaging.

Histopathology In studies of spinal cord injury it is essential to evaluate the pathological changes at the injury site. Furthermore, in developing new models or when using existing models, it is important to correlate the pathological changes at the injury site with the *severity* of the applied injury. With modern image-analysis techniques it is possible to quantify many histopathological changes, which adds greatly to the accuracy of evaluation. For example, some authors have quantified the cross-sectional extent of necrosis (Means *et al.* 1981) and, in our laboratory, we have quantified the extent of haemorrhage of the injury site using image-analysis techniques (Guha, Tator, and Piper 1985*b*).

There are numerous axonal staining techniques which can be utilized as outcome parameters. We have been able to count axons in the rat pyramidal tract after staining with Holmes' silver stain (Tator, Rivlin, Lewis, and Schmoll 1984). Although this can be accomplished at the light microscopic level, electron microscopy should be used to corroborate these counts because of the small size of many of the axons. Indeed, Blight (1983*b*) used an elegant line-sampling technique to count spinal cord axons at both the light and electron microscopic levels in cats subjected to weight-drop injuries.

The Falck–Hillarp histofluorescence technique for visualizing tissue monoamines was introduced in 1962 (Falck, Hillarp, Thieme, and Torp 1962) and has now been largely replaced by the simpler and more sensitive glyoxylic-acid-induced fluorescence technique (Björklund and Skagersberg 1979; de la Torre 1980*b*). This latter technique identifies catecholamine-containing varicosities following cord injury and is used to establish the integrity of catecholaminergic axons across the site of a cord injury (de la Torre 1980*a*). It is important to note that these axonal staining techniques do not assess axonal integrity as sensitively or as specifically as do the axonal tracing techniques described below.

Spinal cord angiography The vasculature of the cord can be imaged radiographically with a contrast medium injected into the aorta or into one of the main arterial feeders of the cord. For example, Allen, d'Angelo, and Kier (1979) used a barium-gel suspension injected into the aortic arch of cats to examine the microangiographic changes in the spinal cord following injury. Another technique used frequently in our laboratory and elsewhere is colloidal carbon angiography (Wallace, Tator, and Frazee 1986), which is performed by injecting colloidal carbon into the abdominal aorta [Fig. 1.4(a)]. In addition to radio-opaque contrast agents and colloidal carbon, other substances have been used to image the cord microvasculature, such as red latex (Allen *et al.* 1979) and fluorescein (Aki and Toya 1984). Since vascular changes in the core are of major importance with respect to recovery and treatment of acute spinal cord injury, these angiographic techniques provide powerful tools for outcome analysis, especially when quantified by image-analysis techniques.

Imaging of the spinal cord Although computerized tomographic (CT) scanning of the spine and spinal cord has become a valuable clinical tool, its inability to resolve the finer anatomical features of the cord has limited its value as a research technique in experimental spinal cord injury (Brant-Zawadzki, Miller, and Federle 1981). In contrast, magnetic resonance imaging (MRI) should be able to resolve soft tissue structure with far greater detail than CT scanning (Modic, Hardy, Weinstein, Duchesneau, Paushter, and Boumphrey 1984). To date, MRI has been used only on a preliminary basis to evaluate the spinal cord anatomy of experimental animals (Hansen 1982). In our laboratory we are planning to use MRI to accurately quantify the amount of haemorrhage and oedema in the cord after trauma. MRI offers the major advantage of non-invasive measurement, thus allowing serial studies to be performed.

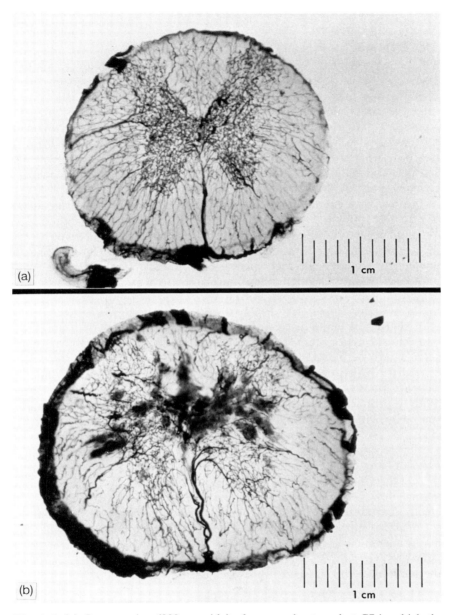

Fig. 1.4. (a) Cross-section (200 μm wide) of a normal rat cord at C7 in which the vessels have been opacified by colloidal carbon angiography. The central grey matter is predominantly supplied by branches of the anterior sulcal artery, and the lateral grey and white matter are supplied by a network of radiating vessels. (b) The microangiographic appearance of a rat cord at C7 following a clip compression injury of 40 g for one minute. Note the haemorrhage in the central grey and dorsal white matter and the ischaemic zones associated with the haemorrhages.

Measurement of SCBF

The measurement of SCBF is of considerable importance in cord injury research. SCBF decreases following cord injury and it has been hypothesized that post-traumatic cord ischaemia enhances tissue destruction after the initial mechanical injury (Sandler and Tator 1976a; Senter and Venes 1979; Guha *et al.* 1985a). Consequently, it has been postulated that an increase in post-traumatic SCBF will minimize core infarction and improve neurological function. Indeed, Guha *et al.* (1985b) have successfully reversed post-traumatic cord ischaemia with the combination of nimodipine, a calcium channel blocker, and adrenalin, a vasopressor.

The two most commonly used quantitative methods of measuring SCBF are the ^{14}C-antipyrine technique and the hydrogen clearance method. The ^{14}C-antipyrine autoradiographic method (Sandler and Tator 1976a) is the most accurate technique for measuring SCBF (Sandler and Tator 1976b). Its advantages include: extremely high resolution, resulting in clear differentiation of grey and white matter flow; measurement of SCBF in the *whole* cord at a given time; avoidance of trauma to the cord at the time of measurement; and provision of a pictorial display of SCBF in histological sections of the entire cord. The major disadvantage of this technique is that only one SCBF determination per animal can be performed since the animal must be sacrificed to derive the measurement.

With the hydrogen clearance technique, originally described by Aukland, Bower, and Berliner (1964), and extensively reviewed by Young (1980), SCBF is proportional to the rate at which inhaled or injected hydrogen gas, sensed by platinum electrodes in the cord, is removed from spinal cord tissue (Fig. 1.5). According to the bicompartmental analysis of clearance curves, the initial slope segment represents the combined grey and white matter flows (Young 1980; Guha *et al.* 1985a). By analysing different segments of the desaturation curve, one can derive separate values for grey and white matter flow. Although this technique necessitates the implantation of electrodes into the cord, we have demonstrated (Guha *et al.* 1985a) that the use of 10 μm microelectrodes does not inflict visible damage to the cord. The major advantage of this technique over the ^{14}C-antipyrine method is that it allows repeated SCBF determinations from multiple sites, and is therefore the method of choice for the measurement of SCBF.

Axonal tracers

Numerous axonal tracers, both anterograde or retrograde, are now available for assessing the integrity of axons in the spinal cord (Chambers 1975). Tracers such as horse-radish peroxidase (HRP) have been extensively used to study normal neuronal connections (Batini, Buisseret-Delmas, Corvisier, Hardy, Jassik-Gerschenseld, and Feld 1978; Mesulam 1978). HRP is the marker most frequently selected for retrograde transport studies since it can

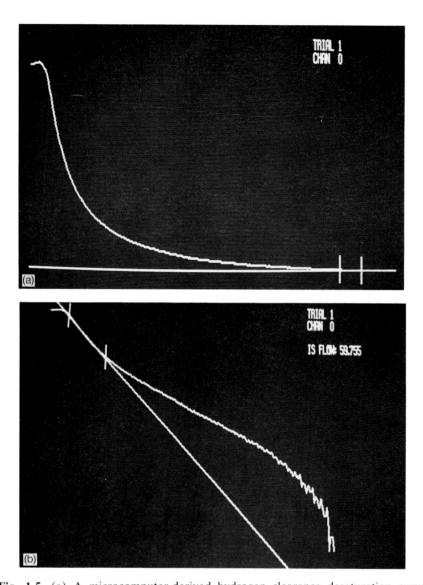

Fig. 1.5. (a) A microcomputer-derived hydrogen clearance desaturation curve measured with platinum/iridium microelectrodes in the cervical cord of a normal rat. The curve is typically bi-exponential, reflecting the grey and white matter flows. (b) The logarithmic transformation of this desaturation curve with a 'best-fit' line plotted through the initial slope segment from one to three minutes. The slope of this line, shown in the upper right corner (IS FLOW = 59.755), represents the combined grey and white matter flow in ml/100 g/min. This type of outcome parameter is quantifiable, objective, and repeatable, and, therefore, is well suited to spinal cord injury experiments.

be readily identified histochemically at both the light and electron microscopic levels (La Vail 1978). Indeed, HRP has also been effectively used to anterogradely label the efferent projections of neurons (La Vail 1978). Recently, axonal tracers, in particular HRP, have been used to study axonal regeneration in the spinal cord. For example, Richardson *et al.* (1978, 1984) have used counts of neurons retrogradely labelled by HRP in the brain stem and spinal cord to assess axonal regeneration through peripheral nerve autografts placed into the cord. In addition to their usefulness in studies of axonal regeneration, axonal tracers have also been used in our laboratory to assess the post-traumatic integrity of specific spinal cord tracts such as the corticospinal and rubrospinal tracts (Midha, Fehlings, Tator, St Cyr, and Guha 1987). These tracer techniques should be considered as essential outcome parameters in studies of axonal regeneration in the spinal cord.

Biochemical evaluation

A useful approach in studying factors involved in the pathophysiology of spinal cord injury is to measure the biochemical alterations in cord tissue following injury. For example, considerable work carried out recently has focused on the potential role of calcium in the pathophysiology of cord injury. It has been shown that total ionic calcium increases, and that of extracellular ionic calcium decreases, at the site of cord trauma (Stokes, Fox, and Hollinden 1983) and that the influx of ionic calcium into the axons of the injured cord appears to have a cytotoxic effect (Balentine and Spector 1977). Other authors have focused on the possible role of catecholamines, in particular noradrenalin, in the pathogenesis of cord injury (Osterholm and Mathews 1972; Schoultz, de Luca, and Reding 1976). Indeed, Jones and McKenna (1980) have recently reported that, in the rat spinal cord, noradrenalin stimulates cyclic AMP formation, a compound which potentiates vasoconstriction of vascular smooth muscle (Hidaka, Yamaki, Totsuka, and Asano 1979). Other compounds which have been assayed following experimental spinal cord injury include Na^+-K^+-ATPase (Clendenon, Allen, Gordon, and Bingham 1978), cytochrome oxidase (Yamada *et al.* 1981), and various energy metabolites including lactate, pyruvate, and adenosine triphosphate (Walker, Yates, O'Neil, and Yashan 1977).

Another very useful biochemical approach involves assessing the overall metabolic activity of the cord. The 2-deoxyglucose autoradiographic technique (Rawe and Perot 1979) is a precise method of quantifying the metabolic activity of the cord and can accurately distinguish between grey and white matter metabolism. Recently, di Chiro, Oldfield, Bairamian, Patronas, Brooks, Mansi, Smith, Kornblith, and Margohn (1983) performed metabolic imaging studies of the brain and spinal cord with a positron emission tomographic (PET) scanner using ^{18}F-2-deoxyglucose. Magnetic resonance spectroscopy has also been used to assess tissue metabolism and in the future may become the preferred technique in non-invasively assessing the metabolic state of the injured cord. For example, the spectroscopic relationship of the inorganic phosphate peak to the phosphocreatine peak is directly related to intracellular pH (Petroff, Prichard, Behar, Alger, den Hollander, and Shulman 1985).

Neurophysiological tests
One of the first neurophysiological tests to be used to assess outcome following spinal cord injury was the H-reflex (Hoffman 1918; d'Angelo, 1973), which has been of only limited value. Since then various evoked potential techniques have been developed to assess the functional integrity of the cord and these have proven to be considerably more useful. Since the first description by Dawson (1947) almost forty years ago, SSEPs have been extensively used to assess the function of the spinal cord clinically (Perot 1973; Nash and Brown 1979) and in experimental animals with cord lesions (Kobrine, Evans, and Rizzoli 1979; Schramm, Shigeno, and Brock 1983; Nacimento, Bartels, and Loew 1985). The SSEP is elicited by stimulating a peripheral nerve and recording the response from the somatosensory cortex. In addition, a spinal-evoked potential can be recorded from the cord. Averaging is used to separate the evoked potential response from background electroencephalographic and other bioelectric activity. Figure 1.6 illustrates the computer-derived grand-mean SSEP obtained from sixteen normal rats following direct sciatic nerve stimulation (Fehlings, Tator, Linden, and Piper 1987). The SSEP is mediated primarily by the dorsal columns (Powers, Bolger, and Edwards 1982; Fehlings *et al.* 1987) and reflects sensory function, but unfortunately it is not a reliable predictor of motor function (York, Watts, Raffensberger, Spagnolia, and Joyce 1983;

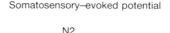

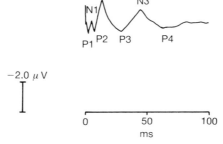

Fig. 1.6. Computer-derived grand-mean SSEP, from sixteen normal rats, following direct sciatic nerve stimulation.

Motor-evoked potential

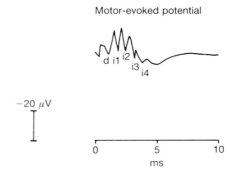

−20 μV

0   5   10
ms

Fig. 1.7. Computer-derived grand-mean MEP, from twenty normal rats, recorded from microelectrodes positioned in the spinal cord at T10. The MEP consists of an initial D-wave, resulting from direct pyramidal cell excitation, and four subsequent i-waves due to indirect pyramidal cell excitation, relayed through cortical interneuron networks.

Ginsburg, Shetter, and Raudzens 1985; Lesser, Raudzens, Luders, Nuwer, Goldie, Morris, Dinner, Klem, Hahn, and Shetter 1986) and should *not* be used to monitor the *motor* tracts of the cord in spinal cord injury experiments.

In an effort to monitor the ventral tracts of the cord, Young, Tomasula, de Crescito, Flamm, and Ransohoff (1980) recorded vestibular-evoked potentials elicited by direct vestibular nerve stimulation in cord-injured cats. Another approach has been to use electromyographic recording techniques to monitor the vestibulospinal free-fall response in cats with spinal cord injuries (Gruner, Young, and de Crescito 1984).

Motor-evoked potentials (MEPs) are the most direct neurophysiological method of monitoring the motor tracts of the spinal cord. In 1983, Levy reported that stimulation of the spinal cord between the intermediolateral *sulcus* and the dentate ligament elicited a MEP, and subsequently he and colleagues recorded MEPs following transcranial stimulation of the motor cortex in cats and humans (Levy, McCaffrey, York, and Tanzer 1984*a*; Levy, McCaffrey, and Tanzer 1984*b*). Fehlings *et al.* (1987) have recently characterized the MEPs recorded from the spinal cord of rats (Fig. 1.7). The MEP consists of an initial positive deflection, the D-wave, which travels with a conduction velocity of 60–70 m/s, and a series of subsequent i-waves. Lesioning studies (Levy *et al.* 1984*a*; Fehlings *et al.* 1987) indicate that the MEP is conducted only by descending motor fibres. Therefore, the MEP method is a clearly superior technique of assessing motor function than either SSEP or vestibulospinal tract monitoring. Furthermore, the combined recording of MEPs and SSEPs monitors *both* the motor and sensory tracts of the cord and hence should be of great value in spinal cord injury research.

Principles of experimental design for studies using models of acute spinal cord injury

Unfortunately, many experiments on spinal cord injury reported in the literature contain serious errors in experimental design. The main flaws have been selection of an inappropriate model and/or species, inadequate sample size, and lack of objectivity. A properly designed experiment is essential in order to derive valid conclusions. While a complete discussion of experimental design is beyond the scope of this chapter, certain principles specific to the use of spinal cord injury models will be emphasized.

Selection of appropriate injury mechanism and species

The model of cord injury selected should be highly reproducible and appropriate for the particular animal species to be studied. Reproducibility implies the delivery of a consistent injury dose to each animal. Selection of an appropriate model for a particular species is often difficult. For example, while the weight-drop model, with the appropriate modifications outlined above, appears to be acceptable for experiments on large animals such as primates or dogs, this model is not recommended for small animals such as the rat (Khan and Griebel 1983). An alternative injury model, such as the clip compression technique, should be used for small animals.

Adequate sample size

An adequate number of animals is required to produce results which are statistically and biologically significant. Statistical techniques for calculating an appropriate sample size are described in standard texts (e.g. Snedecore and Cochran 1980). In most comparison studies at least ten animals per group should be used, although factorial designs can usually operate with fewer animals. Results achieving a significance value of only 0.05 ($p < 0.05$) should be regarded as preliminary and should be repeated as soon as possible, preferably with a larger sample size, and preferably prior to reporting the results in the literature.

Elimination of investigator bias at time of injury

The potential for reward is so great in experiments on spinal cord injury that it is essential to prevent investigator bias. To reduce systematic bias, the animals must be randomized 'blindly' to the treatment groups, including the placebo, *after* the cord injury has been made, otherwise it is impossible to ensure that equal injury doses have been delivered to all groups. Furthermore, the use of historical controls rather than concurrent controls is unacceptable.

Selection of appropriate outcome measures

It is essential to choose an outcome measure that is reproducible, quantifiable, and appropriate to the species being studied. Indeed, the use of more than one outcome measure maximizes the chance of detecting a treatment effect. For example, the combination of clinical assessment of neurological function, evoked potential monitoring, and histological analysis of the injury site is clearly superior to the use of any one of these techniques in isolation.

Elimination of investigator bias at the time of assessment of outcome

It is important to stress that all assessments of outcome following cord injury, including clinical evaluation and histological analysis, must also be conducted 'blindly' without the assessor's knowledge of the experimental groups to which given animals belong.

Statistical analysis

The statistical techniques for data analysis should be established prior to the beginning of an experiment. For example, in comparing the results of treatment in more than two experimental groups, analysis of variance should be used in combination with an appropriate multiple comparison test such as Scheffé's. Multiple t-tests, which result in a high type-1 error (i.e. a high probability of falsely declaring a statistically significant difference) should not be used for this purpose (Snedecore and Cochran 1980).

Overall, the randomized, 'double-blinded', controlled study in which two or more groups are compared is the best choice of design, with rigid adherence to blinding being made both at the time of delivery of the injury and at the time of outcome assessment.

Summary and conclusions

A quantifiable and reproducible model of experimental spinal cord injury is essential for valid, meaningful research in this field. Furthermore, the model must be appropriate for the particular species of animal under investigation. To assess clinically relevant treatments for cord injury, a general cord injury model should be used which simulates human cord injury. Specific models of cord injury, however, may be more appropriate in investigating particular aspects of cord injury pathophysiology such as the role of calcium in neuronal death.

The methods to evaluate outcomes of cord injury include clinical, physiological, biochemical, and anatomical techniques. If clinical evaluation is selected, then a quantifiable and objective method such as the inclined plane technique is essential. Neurological grading schemes based on clinical neurological examination should not be used in subprimate mammals.

Finally, great care should be taken in the experimental design and in the analysis of data. Animals should be randomly allocated to experimental treatments, including the placebo, in a 'blind' manner *after* the cord injury has been inflicted, and outcome assessments must also be performed 'blindly'.

Careful adherence to the above principles will allow the investigator to conduct spinal cord injury experiments which are scientifically sound and biologically significant.

References

Aki, T. and Toya, S. (1984). Experimental study on the circulatory dynamics of the spinal cord by serial fluorescein angiography. *Spine* **9**, 262–7.

Allen, A. R. (1911). Surgery of experimental lesions of spinal cord equivalent to crush injury of fracture dislocation. Preliminary report. *J. Am. Med. Ass.* **57**, 878–80.

Allen, W. F., d'Angelo, C. M., and Kier, F. I. (1979). Correlation of microangiographic and electrophysiologic changes in experimental spinal cord trauma. *Radiology* **111**, 107.

Anderson, D. K., Behbehani, M., Means, E. D., Waters, T., and Green, E. S. (1983). Susceptibility of feline spinal cord energy metabolism to severe incomplete ischemia. *Neurology* **33**, 722–31.

Anderson, T. E. (1982). A controlled pneumatic technique for experimental spinal cord contusion. *J. Neurosci. Meth.* **6**, 327–33.

Ard, M. D., Bunge, R. P., and Bunge, M. B. (1985). Role of Schwann cells and their extracellular matrix in promoting neurite growth in vitro. *Soc. for Neurosci. (Abstract)* **11**(2), 761.

Aukland, K., Bower, B. F., and Berliner, R. W. (1964). Measurement of local blood flow with hydrogen gas. *Circulat. Res.* **14**, 164–87.

Balentine, J. D. and Dean, D. L. (1982). Calcium induced spongiform and nectorizing myelopathy. *Lab. Invest.* **47**, 286–95.

Balentine, J. D. and Spector, M. (1977). Calcification of axons in experimental spinal cord trauma. *Ann. Neurol.* **2**, 520–3.

Banik, N. L., Hogan, E. L., Powers, J. M., and Whetstine, L. J. (1982). Degradation of neurofilaments in spinal cord injury. *Neurochem. Res.* **7**, 1465–75.

Barrett, C. P., Donnati, E. J., and Guth, L. (1981). A comparison of spinal lesions produced by laser beam and conventional scalpel in rats. *Anat. Rec.* **199**, 18A–19A.

Bastian, H. C. (1886). *Paralysis, cerebral, bulbar and spinal.* H. K. Lewis, London.

Batini, C., Buisseret-Delmas, C., Corvisier, J., Hardy, O., Jassik-Gerschenseld, D., and Feld, D. (1978). Brainstem nuclei giving fibers to lobules, VI and II of the cerebellar vermis. *Brain Res.* **153**, 241–61.

Berman, M. L. and Murray, W. J. (1972). Effect of intrathecal epinephrine on rabbit spinal cord. *Anesthesia and Analgesia* **51**, 383–6.

Björklund, A. and Skagerberg, G. (1979). Simultaneous use of retrograde fluorescent tracers and fluorescences histochemistry for convenient and precise mapping of monoaminergic projections and collateral arrangements in the CNS. *J. Neurosci. Meth.* **1**, 261–77.

Blight, A. R. (1983a). Axonal physiology of chronic spinal cord injury in the cat: intracellular recording in vitro. *Neuroscience* **10**, 1471–86.

Blight, A. R. (1983b). Cellular morphology of chronic spinal cord injury in the cat: analysis of myelinated axons by line-sampling. *Neuroscience* **10**, 521–43.

Blight, A. R. (1985). CNS injury: acute response, recovery, and repair. Paper presented at the Third Annual Neural Trauma Symposium of the Soc. for Neurosci., 15th Annual Meeting, Dallas, Texas, 20 October.

Borgens, R. B., Roederer, E., and Cohen, M. J. (1981). Enhanced spinal cord regeneration in lamprey by applied electric fields. *Science* **213**, 611–17.

Brant-Zawadzki, M., Miller, E. M., and Federle, M. P. (1981). CT in the evaluation of spine trauma. *Am. J. Roent.* **136**, 369–75.

Breasted, J. H. (1930). *The Edwin Smith surgical papyrus*, vol. 1, pp. 337–42. University of Chicago Press, Chicago, Illinois.

Bresnahan, J. C., Todd, F. D., Noyes, D. H., and Beattie, M. S. (1985). A behavioural and anatomical analysis of spinal cord injury produced by a feedback controlled computerized impaction device. *Soc. for Neurosci. (Abstract)* **11**(2), 1167.

Bunge, R. P., Johnson, M., and Thuline, D. (1983). Spinal cord reconstruction using embryonic spinal strips. In *Spinal cord reconstruction* (ed. C. Kao, R. Bunge, and P. Reir), pp. 341–57. Raven Press, New York.

Chambers, A. M. (1975). Wallerian degeneration and anterograde tracer methods. In *The use of axonal transport for studies of neuronal connectivity* (ed. W. M. Cowan and M. Cuerod), pp. 174–216. Elsevier, New York.

Clendenon, N. R., Allen, N., Gordon, W., and Bingham, G. (1978). Inhibition of Na+-K+-activated ATPASE activity following experimental spinal cord trauma. *J. Neurosurg.* **49**, 563–8.

Coughlin, M. D., Grover, A. K., and Jung, C. Y. (1985). Comparison of laminin and neuronectin (a conditioned medium-derived neurite extension factor) using radiation inactivation analysis. *Soc. for Neurosci. (Abstract)* **11**(1), 175.

Croft, T. J., Brodkey, J., and Nulsen, F. E. (1972). Reversible spinal cord trauma: a model for electrical monitoring of spinal cord function. *J. Neurosurg.* **36**, 402–6.

Cusick, J. F., Myklebust, J., Zyvoloski, M., Sances, A., Houterman, C., and Larson, S. J. (1982). Effects of vertebral column distraction in the monkey. *J. Neurosurg.* **57**, 651–9.

D'Angelo, C. M. (1973). The H-reflex in experimental spinal cord trauma. *J. Neurosurg.* **39**, 209–13.

David, S. and Aguayo, A. J. (1981). Axonal elongation into peripheral nervous system 'bridges' after central nervous system injury in adult rats. *Science* **214**, 931–3.

Dawson, G. D. (1947). Investigations in a patient subjected to myoclonic seizures after sensory stimulation. *J. Neurol. Neurosurg. Psych.* **10**, 141–62.

De la Torre, J. C. (1980a) Chemotherapy of spinal cord trauma. In *The spinal cord and its reaction to traumatic injury* (ed. W. Windle), pp. 291–310. Marcel Dekker, New York.

De la Torre, J. C. (1980*b*). An improved approach to histofluorescence using the SPG method for tissue monoamines. *J. Neurosci. Meth.* **3**, 1–5.

Derlon, J. M., Camille-Roy, R., Lechevalier, B., Bisserie, M., and Coston, A. (1983). Delayed spinal cord anastomosis. In *Spinal cord reconstruction* (ed. C. C. Kao, R. P. Bunge, and P. Reir), pp. 223–34. Raven Press, New York.

Di Chiro, G., Oldfield, E., Bairamian, D., Patronas, N., Brooks, R., Mansi, L., Smith, B. H., Kornblith, P. L., and Margohn, R. (1983). Metabolic imaging of the brainstem and spinal cord: studies with PET using F-2-deoxyglucose in normal and pathologic cases. *J. Comp. Asst. Tom.* **7**, 937–45.

Dohrmann, G. J. and Panjabi, M. M. (1976). 'Standardized' spinal cord trauma: biomechanical parameters and lesion volume. *Surg. Neurol.* **6**, 263–7.

Dohrmann, G. J., Panjabi, M. M., and Banks, D. (1978). Biomechanics of experimental spinal cord trauma. *J. Neurosurg.* **48**, 993–1001.

Dolan, E. J. and Tator, C. H. (1979). A new method for testing the force of clips for aneurysms of experimental spinal cord compression. *J. Neurosurg.* **51**, 229–33.

Dolan, E. J., Tator, C. H. and Endrenyi, L. (1980). The value of decompression for acute experimental spinal cord compression injury. *J. Neurosurg.* **53**, 749–55.

Dolan, E. J., Transfeldt, E. E., Tator, C. H., Simmons, E. H., and Hughes, K. (1980). The effect of spinal distraction on regional spinal cord blood flow in cats. *J. Neurosurg.* **53**, 756–64.

Doppman, J. L., Girton, M., and Popovsky, M. A. (1979). Acute occlusion of the spinal vein. Experimental study in monkeys. *J. Neurosurg.* **51**, 201–4.

Edwards, M. S., Boggan, J. E., and Fuller, T. A. (1983). The laser in neurological surgery. *J. Neurosurg.* **59**, 555–6.

Eidelberg, E., Staten, E., Watkins, J. C., McGraw, D., and McFadden, C. (1976). A model of spinal cord injury. *Surg. Neurol.* **5**, 35–8.

Falck, B., Hillarp, N. A., Thieme, G., and Torp, A. (1962). Fluorescence of catecholamines and related compounds condensed with formaldehyde. *J. Histochem. Cytochem.* **10**, 348–54.

Fehlings, M. G., Tator, C. H., Linden, R. D., and Piper, I. R. (1988). Motor and somatosensory evoked potentials recorded from the rat. *Electroenceph. Clin. Neurophysiol.* **69**, 65–78.

Fialho, S. A. G., Lumb, W. V., and Scott, R. J. (1982). Pneumatically powered vertebral displacement device for dogs. *Am. J. Vet. Res.* **43**, 1254–7.

Ford, R. J. W. (1983). A reproducible spinal cord injury model in the cat. *J. Neurosurg.* **59**, 269–75.

Freeman, L. W. and Wright, T. W. (1953). Experimental observations of concussion and contusion of the spinal cord. *Ann. Surg.* **137**, 433–40.

Fried, L. C., di Chiro, G., and Doppman, J. L. (1969). Ligation of major thoraco-lumbar spinal cord arteries in monkeys. *J. Neurosurg.* **31**, 608–14.

Ginsburg, H. H., Shetter A. G., and Raudzens, D. A. (1985). Post-operative paraplegia with preserved intraoperative somatosensory evoked potentials. A case report. *J. Neurosurg.* **63**, 296–300.

Grillner, S. (1973). Locomotion in the spinal cat. In *Control of posture and locomotion* (ed. R. B. Stein), pp. 515–35. Plenum Press, New York.

Grillner, S. (1975). Locomotion in vertebrates: central mechanisms and reflex interaction. *Physiol. Rev.* **55**, 247–304.

Gruner, J. A., Young, W., and de Crescito, V. (1984). The vestibulo-spinal free fall response: a test of descending function in spinal-injured cats. *CNS Trauma* **1**, 139–59.

Guha, A., Tator, C. H., and Piper, I. (1985*a*). Increase in rat spinal cord blood flow with the calcium channel blocker, nimodipine. *J. Neurosurg.* **63**, 250–9.

Guha, A., Tator, C. H., and Piper, I. (1985*b*). Increase in posttraumatic spinal cord blood flow with a calcium channel blocker. *Can. J. Neurol. Sci.* **12**, 195.

Guth, L., Brewer, C. R., Collins, W. F., Goldberger, M. E., and Perl, E. R. (1980). Criteria for evaluating spinal cord regeneration in experiments. *Exp. Neurol.* **69**, 1–3.

Hansen, G. (1982). In vivo, imaging of the rat anatomy with nuclear magnetic resonance. *Radiology* **136**, 695–700.

Hidaka, H., Yamaki, T., Totsuka, T., and Asano, M. (1979). Selective inhibitors of Ca^{2+}-binding modulator of phosphodiesterase produce vascular relaxation and inhibit actin–myosin interaction. *Mol. Pharmacol.* **15**(1), 49–59.

Hoffman, P. (1918). Über die Beziehungen der Sehnenreflexe zur wilkurlichen Bewegung und zum Tonus. *Z. Biol.* **68**, 351–70.

Hukuda, S. and Wilson, C. B. (1972). Experimental cervical myelopathy: effects of compression and ischemia on the canine cervical cord. *J. Neurosurg.* **37**, 631–52.

Hukuda, S., Mochizuki, T., and Ogata, M. (1980). Effects of hypertension and hypercarbia on spinal cord tissue oxygen in acute experimental spinal cord injury. *Neurosurgery* **37**, 639.

Jackson, M. B., Lecar, H., Brenneman, D. E., Fitzgerald, S., and Nelson, P. (1982). Electrical development in spinal cord cell culture. *J. Neurosci.* **2**, 1052–61.

Jones, D. J. and McKenna, L. F. (1980). Norepinephrine-stimulated cyclic AMP formation in rat spinal cord. *J. Neurochem.* **34**, 467–9.

Kao, C. C. (1974). Comparison of healing process in transected spinal cords grafted with autogenous brain tissue, sciatic nerve and nodose ganglion. *Exp. Neurol.* **44**, 424–39.

Khan, M. and Griebel, R. (1983). Acute spinal cord injury in the rat: comparison of three experimental techniques. *Can. J. Neurosci.* **10**, 161–5.

Khan, M., Griebel, R., Rozdilsky, B., and Politis, M. (1985). Hemorrhagic changes in experimental spinal cord injury models. *Can. J. Neurol. Sci.* **12**, 259–62.

Kobrine, A. I., Evans, D. E., and Rizzoli, H. V. (1979). The effects of ischemia on long-tract neural conduction in the spinal cord. *J. Neurosurg.* **50**, 639–44.

Koozekanani, S. H., Vise, M., Hashemi, R., and McGhee, R. (1976). Possible mechanisms for observed pathophysiological variability in experimental spinal cord injury by the method of Allen. *J. Neurosurg.* **44**, 429–34.

Kraus, J. F., Franti, C. E., Riggins, R. S., Richards, D., and Borhani, N. O. (1975). Incidence of traumatic spinal cord lesions. *J. Chron. Dis.* **28**, 471–92.

Krogh, E. (1950). The effect of acute hypoxia on the motor cells of the spinal cord. *Acta Physiol. Scand.* **20**, 263–92.

La Vail, J. H. (1978). A review of the retrograde transport technique. In *Neuro-anatomical research techniques* (ed. R. T. Robertson), pp. 355–84. Academic Press, New York.

Lesser, R. P., Raudzens, P., Luders, H., Nuwer, M. R., Goldie, W. D., Morris, H. H., Dinner, D. S., Klem, G. K., Hahn, J. F., and Shetter, A. G. (1986). Postoperative neurological deficits may occur despite unchanged intraoperative somatosensory evoked potentials. *Ann. Neurol.* **19**, 22–5.

Levy, W. J. (1983). Spinal evoked potentials from the motor tracts. *J. Neurosurg.* **58**, 38–44.

Levy, W. J., McCaffrey, M., and Tanzer, F. (1984*b*). Motor evoked potentials from transcranial stimulation of the motor cortex in humans. *Neurosurgery* **15**, 287–302.

Levy, W. J., McCaffrey, M., York, D. H., and Tanzer, F. (1984*a*). Motor evoked potentials from transcranial stimulation of the motor cortex in cats. *Neurosurgery* **15**, 214–27.

Madison, R., da Silva, C. F., Dikkes, P., Chui, T. H., and Sidman, R. L. (1985). Increased rate of peripheral nerve regeneration using bioresorbable nerve guides and a laminin-containing gel. *Exp. Neurol.* **88**, 767–72.

McCallum, H. E. and Bennett, M. H. (1976). DMSO as a therapeutic agent in chronic spinal cord compression. In *Program abstracts of the 26th Annual Meeting of the Congress of Neurological Surgeons, New Orleans*, pp. 166–7. Congress of Neurological Surgeons.

McVeigh, J. F. (1923). Experimental cord crushes with especial reference to the mechanical factors involved and subsequent changes in the areas of the cord affected. *Arch. Surg.* **7**, 573–600.

Means, E. D., Anderson, D. K., Waters, T. R., and Kalaf, L. (1981). Effect of methylprednisolone in compression trauma to the feline spinal cord. *J. Neurosurg.* **55**, 200–8.

Mesulam, M. A. (1978). Tetramethylbenzidine for horseradish peroxidase neurohistochemistry. A non-carcinogenic blue reaction product with superior sensitivity for visualizing neural afferents and efferents. *J. Histochem. Cytochem.* **26**, 106.

Midha, R., Fehlings, M. G., Tator, C. H., St Cyr, J., and Guha, A. (1987). Assessment of spinal cord injury by counting corticospinal and rubrospinal neurons. *Brain Res.* **410**, 299–308.

Modic, M. T., Hardy, R. W. Jr., Weinstein, M. A., Duchesneau, P. M., Paushter, D. M., and Boumphrey, F. (1984). Nuclear magnetic resonance of the spine: clinical potential and limitation. *Neurosurgery* **15**, 583–92.

Naciemento, A. C., Bartels, M. M., and Loew, F. (1985). Acute changes in somatosensory evoked potentials following graded experimental spinal cord compression. *Surg. Neurol.* **25**, 62–6.

Nash, C. L. and Brown, R. H. (1979). The intraoperative monitoring of spinal cord function: its growth and current status. *Orth. Clin. N. Amer.* **1**, 919–26.

Nobin, A., Baumgarten, H. G., Björklund, A., Lachenmayer, L., and Stenevi, U. (1973). Axonal degeneration and regeneration of the bulbo-spinal indoleamine neurons after 5,6-dihydroxytryptamine treatment. *Brain Res.* **56**, 1–24.

Nornes, H., Björklund, A., and Stenevi, U. (1983). Reinnervation of the denervated adult spinal cord of rats by intraspinal transplants of embryonic brain stem neurons. *Cell Tissue Res.* **230**, 15–35.

Osterholm, J. L. and Mathews, G. J. (1972). Altered norepinephrine metabolism following experimental spinal cord injury. Part 1: relationships to hemorrhagic necrosis and post-wound deficits. Part 2: protection against traumatic spinal cord hemorrhagic necrosis by norepinephrine by synthesis blockage with alphamethyltyrosine. *J. Neurosurg.* **20**, 382–99.

Patel, N. and Poo, M. M. (1982). Orientation of neurite growth by extracellular electric fields. *J. Neurosci.* **2**, 483–96.

Patel, N. and Poo, M. M. (1984). Purturbation of the direction of neurite growth by pulsed and focal electric fields. *J. Neurosci.* **4**, 2939–47.

Perot, P. L. Jr. (1973). The clinical use of somatosensory evoked potentials in spinal cord injury. *Clin. Neurosurg.* **20**, 367–81.

Petroff, O. A. C., Prichard, J. W., Behar, K. L., Alger, J. R., den Hollander, J. A., and Shulman, R. G. (1985). Cerebral intracellular pH by 31P nuclear magnetic resonance spectroscopy. *Neurology* **35**, 781–8.

Powers, S. K., Bolger, C. A., and Edwards, M. S. B. (1982). Spinal cord pathways mediating somatosensory evoked potentials. *J. Neurosurg.* **57**, 472–82.

Rawe, S. E. and Perot, P. L. (1979). Autoradiographic technique for the study of metabolism in experimental spinal cord injury. In *Neural trauma* (ed. A. J. Popps), pp. 35–9. Raven Press, New York.

Richardson, P. M., Issa, V. M. K., and Aguayo, A. J. (1984). Regeneration of long spinal axons in the rat. *J. Neurocyt.* **13**, 165–82.

Richardson, P. M., McGuiness, U. M., and Aguayo, A. J. (1980). Axons from CNS neurones regenerate into PNS grafts. *Nature* **284**, 264–5.

Richardson, P. M., McGuiness, U. M., and Aguayo, A. J. (1982). Peripheral nerve autografts to the rat spinal cord studies with axonal tracing methods. *Brain Res.* **237**, 147–62.

Riese, W. (1959). *A history of neurology*, p. 223. MD Publications, New York.

Rivlin, A. S. and Tator, C. H. (1977). Objective clinical assessment of motor function after experimental spinal cord injury in the rat. *J. Neurosurg.* **47**, 577–81.

Rivlin, A. S. and Tator, C. H. (1978a). Effect of duration of acute spinal cord compression in a new acute cord injury model in the rat. *Surg. Neurol.* **9**, 39–43.

Rivlin, S. S. and Tator, C. H. (1978b). Regional spinal cord blood flow in rats, after severe cord trauma. *J. Neurosurg.* **49**, 844–7.

Ross, R. T. (1985). Spinal cord infarction in disease and surgery of the aorta. *Can. J. Neurol. Sci.* **12**, 289–95.

Rudin, D. O. and Eisenman, G. (1951). A method for dissection and electrical study in vitro of mammalian central nervous tissue. *Science* **114**, 300–2.

Sandler, A. N. and Tator, C. J. (1976a). Effect of acute spinal cord compression injury on regional spinal cord blood flow in primates. *J. Neurosurg.* **45**, 660–76.

Sandler, A. N. and Tator, C. H. (1976b). Review of the measurement of spinal cord blood flow. *Brain Res.* **118**, 181–98.

Schmaus, H. (1890). Commotio spinalis. In *Ergebnisse der allemeinen pathologie und pathologischen Anatomie des Menschen und der tiere* (ed. O. Lubarsch and R. Ostertag), pp. 674–713. J. F. Bergman, Wiesbaden.

Schoultz, T. W., de Luca, D. C., and Reding, D. L. (1976). Norepinephrine levels in traumatized spinal cord of catecholamine-depleted cats. *Brain Res.* **109**, 367–79.

Schramm, J., Shigeno, T., and Brock, M. (1983). Clinical signs and evoked response alterations associated with chronic experimental cord compression. *J. Neurosurg.* **58**, 734–41.

Senter, H. and Venes, J. (1979). Loss of autoregulation and post-traumatic ischemia following experimental spinal cord injury. *J. Neurosurg.* **50**, 198–206.

Siegel, R. E. (1973). *Galen on psychology, psychopathology and function and diseases of the nervous system*, p. 310. Karger, Basel.

Snedecore, G. W. and Cochran, W. G. (1980). *Statistical methods*, 7th edition. Iowa State University Press, Ames, Iowa.

Spiller, W. G. (1899). A critical summary of recent literature on concussion of the spinal cord with some original observations. *Am. J. Med. Sci.* **118**, 190–8.

Stokes, B. T., Fox, P., and Hollinden, G. (1983). Extracellular calcium activity in the injured spinal cord. *Exp. Neurol.* **80**, 561–72.

Tarlov, I. M. (1957). *Spinal cord compression: mechanism of paralysis and treatment.* Thomas, Springfield, Illinois.

Tator, C. H. (1972). Acute spinal cord injury in primates produced by an inflatable extradural cuff. *Can. J. Surg.* **16**, 222–31.

Tator, C. H. (1983). Spine–spinal cord relationships in spinal cord trauma. *Clin. Neurosurg.* **30**, 479–94.

Tator, C. H. and Edmonds, V. E. (1979). Acute spinal cord injury: analysis of epidemiologic factors. *Can. J. Surg.* **22**, 575–8.

Tator, C. H. and Rowed, D. W. (1979). Current concepts in the immediate management of acute spinal cord injuries. *Can. Med. Ass. J.* **121**, 1453–64.

Tator, C. H., Rivlin, A. S., Lewis, A. J., and Schmoll, B. (1984). Effect of acute spinal cord injury on axonal counts in the pyramidal tract of rats. *J. Neurosurg.* **61**, 118–23.

Unterharnscheidt, F. (1983). Neuropathology of rhesus monkeys undergoing Gx impact acceleration. In *Impact injury of the head and spine* (ed. C. Weing, D. Thomas, A. Sances, and S. Larson), pp. 94–176. Charles C. Thomas, Springfield, Illinois.

Ushio, Y., Posner, R., Kim, J. H., Shapiro, W. R., and Posner, J. (1977). Treatment of experimental spinal cord compression caused by extradural neoplasms. *J. Neurosurg.* **47**, 380–90.

Walker, J. G., Yates, R. R., O'Neil, J. J., and Yashan, D. (1977). Canine spinal cord energy state after experimental trauma. *J. Neurochem.* **29**, 929–32.

Wallace, M. C., Tator, C. H., and Frazee, P. (1986). The relationship between posttraumatic ischemia and hemorrhage in the injured rat spinal cord as shown by colloidal carbon angiography. *Neurosurgery* **18**, 433–9.

Wallace, M. C., Tator, C. H., and Piper, I. (1984). The effect of epidural direct current stimulation on recovery following spinal cord injury. *Can. J. Neurol. Sci.* **11**, 287.

Wilson, D. H. (1984). Peripheral nerve implants in the spinal cord in experimental animals. *Paraplegia* **22**, 230–7.

Woollam, D. H. M. and Millen, J. W. (1955). The arterial supply of the spinal cord and its significance. *J. Neurol. Neurosurg. Psychiat.* **18**, 97–102.

Yamada, S., Knierim, D., Maeda, G., and Schulz, R. (1981). Stretch trauma of the spinal cord—oxidative metabolism and ultrastructural changes. *Soc. Neurosci.* **7**, 612.

York, D. H., Watts, C., Raffensberger, M., Spagnolia, T., and Joyce, C. (1983). Utilization of somatosensory evoked cortical potentials in spinal cord injury. Prognostic limitations. *Spine* **8**, 832–9.

Young, W. (1980). H_2 clearance measurement of blood flow: a review of technique and polarographic principles. *Stroke* **11**, 552–64.

Young, W., Tomasula, J., de Crescito, V., Flamm, E. S., and Ransohoff, J. (1980). Vestibulospinal monitoring in experimental spinal trauma. *J. Neurosurg.* **52**, 64–72.

Zivin, J. A. and DeGirolami, V. (1980). Spinal cord infarction: a highly reproducible stroke model. *Stroke* **11**, 200–4.

2

Pathological changes after spinal cord injury

J. T. Hughes

Treatment and management of spinal cord injuries should be based on a scientific evaluation of the therapeutic problem, and so needs to take account in detail of the pathological states found in the injured spinal cord. As I shall relate, these states are not static but form a continuous pageant of pathological change. It is convenient to consider these stages under three 'time' headings, early, intermediate, and late pathological changes after the precipitating spinal cord injury (Hughes 1978).

Early pathological changes

The immediate pattern of damage in a spinal cord injury depends on the nature of the damaging force, and will differ appreciably in a crushing injury from, for example, a discrete penetrating wound. Here I shall consider the common form of serious spinal injury arriving at a paraplegic centre, with gross injury, such as that sustained in a road traffic accident or similar gross trauma where there is extensive bony and soft tissue damage. This common type of injury is often described briefly by the term fracture–dislocation of the spine.

The various tissues damaged may be subdivided into: neuron cell bodies, nerve fibres, neuroglia, meninges, and connective tissues.

The neuron cell bodies are destroyed either directly by the trauma or indirectly by an effect on the vascular supply to the cord. This latter mechanism of damage is of importance, and particularly in incomplete lesions since in these cases some difference in management or therapy may improve outcome. The major blood vessels may be damaged directly by trauma and may be affected by spasm (Hughes 1981). A common phenomenon which occurs some hours after trauma is swelling of the spinal cord which then begins to compress itself within the restricting dura and leptomeninges. Measures to combat swelling are difficult to apply success-fully but would have an important beneficial effect in increasing the degree of recovery in incomplete lesions. These vascular changes and this problem of swelling will be discussed in more detail below.

The consequence of the loss of these destroyed neuron cell bodies depends on the magnitude and site of the destruction. If, for example, the cervical enlargement is destroyed there is a profound lower motor neuron paralysis added to the effects of the spinal cord transection. In contrast, if the mid-thoracic cord is the site of damage there may be a complete cord transection but the lower motor neuron effect is trivial.

The all-important spinal cord nerve fibres, with which we are greatly concerned, are of several types, but a major component are those which form the long tracts in the three white columns. These fibres are damaged directly and also are susceptible to the effects of ischaemia and of swelling, although they are less affected than the neuron cell bodies. The nerve fibres undergo at first swelling and then in severe lesions disintegration of both axonic material and of myelin. The axons, when mildly damaged, show beading, but a more severe state is seen as a line of droplets of axonic material in the former position of the axon, whilst in grossly contused specimens appropriate methods will demonstrate a silver-impregnated dust of displaced axonic fragments. The place where an axon is completely torn may develop a terminal swelling called an end-bulb. The myelin tubes show a corresponding series of changes ranging from swelling of the myelin sheath, seen, in myelin stains under high magnification, as a vesicular appearance, to complete fragmentation of the myelin giving abundant fatty droplets scattered amid the axonic material.

The effect on the neuroglia is dependent on the degree of the damage. In gross trauma the macroglia (astroglia and oligodendroglia) completely disappear. In lesser trauma both may survive to some degree, as we shall see when I consider the intermediate and late stages after spinal cord injury. The role of the microglia is quite different. Those cells of microglia present locally are joined by numerous macrophages (described in the next section).

The meninges are much more resistant to trauma, and ischaemia affects this type of tissue much less than the more vulnerable neurons. The meninges have a very important effect in determining the containment of the swollen cord. I shall return to this factor after I have described the changes in the connective tissues.

In the connective tissue changes are seen the various components of the exudative tissue changes of the damaged cord. The exudative changes are oedema, red-cell diapedesis, and a prominent inflammatory reaction of polymorphs, lymphocytes, and plasma cells. These changes are best seen a little away from the main centre of injury, where they are obscured by the haemorrhage and haemorrhagic necrosis caused directly by the trauma. The oedema, if slight, is seen as an expansion of the perivascular space and as an enlargement of the perineuronal space. If the oedema is severe, it affects the whole cross-sectional area of the cord. All these exudative changes cause swelling of the spinal cord, which becomes rounded and tense within the leptomeninges and dura, with virtual obliteration of the subarachnoid and subdural spaces. In addition to the appearances of swelling there is purplish discoloration, arising initially from the haemorrhage and haemorrhagic necrosis and later from the venous stasis, itself secondary to the swelling. This swollen oedematous region of traumatized spinal cord is very striking and its shape has aroused much interest. It seems clear that this type of cord swelling extends the region of spinal cord destruction and the final form is in the shape of a spindle. This consists of a fusiform region of cord softening affecting several consecutive spinal cord segments and ending above and below by tapering to end in a small area, round in cross-section, situated usually in the posterior columns. This round area, as seen in a transverse section, has frequently been described, and sometimes without the author being aware that it forms part of a larger area in the form of a spindle, a feature evident in serial sections. The round area is a core of damaged tissue forced up by the pressure of the swelling. The final appearances to the pathologist may be exaggerated by post-mortem handling, which may also force softened cord material into the spinal root dural sleeves.

It is tempting to resort to measures to combat this dramatic spinal cord swelling, but the solution to this problem has so far defied most therapeutic approaches. Pharmacological measures to shrink swollen brain tissue fail in just those cases, of head injury, tumour, and abscess, in which a reduction of volume would be most desirable. These measures have been tried in cases of spinal cord injury probably without success, although it is most difficult to evaluate the effects of therapy in such cases. As the swollen spinal cord is contained and restrained by the tight meninges it might seem that the surgical operation of laminectomy with a posterior longitudinal incision of the dura, and possibly also of the pia-arachnoid, might successfully relieve the pressure within the spinal cord. This procedure has been tried but not, unfortunately, with a beneficial result. What happens is that the spinal cord herniates through the meningeal incision, adding, by the consequent move-ment and distortion of the softened cord, to the spinal cord damage. A similar problem is well known in cerebral swelling, which also cannot be decompressed by craniectomy and dural incision without causing further and substantial damage to the brain.

We now come to a consideration of the major vessels of the spinal cord and their reaction to injury. The major spinal arteries are obviously of importance, but even in gross trauma they are seldom severed or completely occluded. The very important single anterior spinal artery and the two variable posterior spinal arteries can usually be found, unoccluded, in the traumatic area, even in cases where the whole parenchyma of the spinal cord is destroyed. It is possible, but unlikely, that these arteries are badly damaged but are restored by repair. A phenomenon which is of importance and for which there is some indirect evidence in spinal cord injury is that these major arteries are affected by spasm (Hughes 1981). This phenomenon is known in head injuries and is probably seen throughout most tissues of the body as a reaction to injury. For example, in traumatic limb injuries a severed major artery will close by spasm and thus stop loss of blood which might be fatal. In the context of spinal cord trauma, arterial spasm seems to have the effect of causing ischaemia and further damage to the spinal cord by swelling. If we develop measures to prevent or abolish this spasm then it might be at the cost of encouraging more bleeding into the traumatic area. The small arteries and capillaries have the important changes referred to above in the exudative phenomena causing the swollen cord. The venous drainage is very abnormal in the acute stage of cord trauma, when the veins are enlarged and engorged with blood. This is a temporary state of enlargement and hyperanaemia, not usually accompanied by thrombosis, unless there is the complication of sepsis. Veins are much more malleable than arteries, and a reconstituted and remodified venous drainage eventually arises. I have previously commented on the obstructive effect on the venous drainage of the swollen cord within the meninges.

Intermediate pathological changes

The pathological changes described above are those seen in the injured spinal cord in the immediate post-traumatic period. After two to three weeks, these acute changes subside and are gradually replaced by a series of reparative changes of prolonged duration, which may extend for two or more years. These changes have such important differences from those seen in the early and late stages that they merit a separate description.

The oedema has now largely subsided and the smaller haemorrhages and area of haemorrhagic necrosis have now been absorbed. The cases with large areas of haemorrhage and haemorrhagic necrosis form cysts and these, if of considerable longitudinal extent, have the form of a tube or syrinx. The development of this syrinx into progressive syringomyelia will be described later. In this intermediate stage, the polymorphs of the acute inflammatory reaction are replaced by lymphocytes and macrophages. The most abundant and striking cell everywhere is the lipid phagocyte, also known as the fat granule cell or compound granular corpuscle. This type of phagocyte is present wherever there is destruction by necrosis of nerve fibres and their myelin. Whilst the breakdown of myelin contributes the major part of the content of these lipid phagocytes, there are also the remnants of destroyed neuron cell bodies and glia. The lipid phagocytes cluster around small vessels forming rosette-like aggregates. In grossly damaged areas of the spinal cord there are innumerable closely packed lipid phagocytes. Where the damage is slight, as in the marginal areas of the main focus of trauma, only a few scattered fat granule cells are seen. The presence and amount of astrocytic gliosis in any particular part of a lesion depends on the degree of damage. Where the damage is slight, there is a reactive astrocytic gliosis. In severely damaged areas the glia has been destroyed with the parenchyma of the spinal cord and so organization and repair is by the formation of connective tissues from fibroblasts and not by the formation of glial fibres from astrocytes. Oligodendrocytes are destroyed or damaged by the trauma. If the damage is slight then some regrowth of the oligodendrocyte and its myelin is possible. The neuronal changes seen in this intermediate stage consist only of a few examples of central chromatolysis (axonal reaction), and these changes, of swollen cell cytoplasm and eccentric nuclei, persist for several years. Gradually over several years this intermediate picture is replaced by the last stage to be described.

Late pathological changes

When the survival period after injury is more than five years the changes found at necropsy are different again to a degree that depends on this lengthening interval. There is now an organizing connective tissue scar with predominately acellular collagenous connective tissue uniting the pachymeninges to the damaged spinal canal and to the scarred spinal cord. In this traumatic scar the anatomical tissue planes may be quite indistinct.

The histological features are seen best in a connective tissue stain such as Van Gieson. The grossly damaged regions of the spinal cord are replaced by connective tissue occupying the site of the former blood clot and haemorrhagic spinal cord necrosis. The less-damaged spinal cord, well seen above and below the main traumatic area, shows an intense astrocytic fibrous gliosis. Here there is a mossy feltwork of proliferated glial fibres arising from a few very large glial cell bodies and replacing the lost nerve fibres and neuron cell bodies. An interesting regenerative phenomenon is often seen in these traumatic scars (Hughes 1984). The cell bodies of the intact posterior root ganglia regrow their central severed nerve processes into the region of the connective tissue scar. These nerve fibres, which are myelinated, but enfolded with peripheral nerve myelin from Schwann cells, grow in a tangled skein in the connective tissue of the scar. They ramify here like the tangle of fibres seen in an amputation neuroma. They do not enter the neuroglial scar and do not enter the intact spinal cord. Consequently the regrowing nerve fibres never make contact with other neurons and thus no physiological recovery occurs.

Traumatic syringomyelia

I now come to consider the interesting and important sequel of syringomyelia after spinal cord trauma. I have described earlier how in the early stage after injury there is in the central part of the spinal cord an area of haemorrhagic necrosis sometimes with blood clots and how in this stage it can form a longitudinal spindle-shaped region of necrosis. In the intermediate stage after injury, this region becomes filled with lipid phagocytes. The common sequence then is the gradual resorption of this softened tissue to form a scar made up of connective tissue in the badly damaged areas and of glia in those areas where the damage was less severe.

Sometimes, however, there is a different sequel. Long after the initial trauma and at a time when the neurological state has been static for several months there are symptoms and signs of an extension upwards of the neurological level of the spinal cord deficit. This extension is due to the formation of an actual cavity, which, having a considerable longitudinal extent, has the shape of a tube, and consequently is called a syrinx. The syrinx may progress relentlessly upwards to reach the medulla. The symptoms and signs are similar to those seen in idiopathic syringomyelia. In the majority of cases the syrinx is an upward extension from the site of trauma, but, in other cases, the exact number is difficult to identify and there is a downward extension. I have also examined cases at autopsy in which both upward- and downward-directed syringes are present in the same individual. The downward-progressing syrinx is difficult or even impossible to recognize clinically, since the cord lesion above has already caused paralysis and sensory loss below.

The mechanism of formation of these syringes is now clear from an examination of the fluid dynamics of the structures found at autopsy. There is always at some stage a communication between the syrinx and the subarachnoid space and this is usually at the posterior root entry zone. There may be more than one communication, and also the site and number of communications may vary during the course of the patient's post-traumatic progress of ascending paraplegia. The protein content of the fluid within the syrinx compared with the protein content in the CSF of the lumbar theca gives an indication of the likelihood of communication. Why the syrinx arises and enlarges is explained by the formation of the communication and by the passage into the syrinx of fluid from the CSF. A valve-like mechanism with a pressure gradient into the syrinx from the subarachnoid space is postulated and probably occurs although the evidence for this is difficult to obtain.

The recognition of this syndrome and the elucidation of the hydrodynamics of the syrinx has been followed by the development of several surgical procedures to arrest the upward extension of the cavity. This is a valuable therapeutic resource to combat a very troublesome complication in a paraplegic.

References

Hughes, J. T. (1978). *Pathology of the spinal cord* (2nd edition). Lloyd-Luke, London.

Hughes, J. T. (1981). Spinal cord trauma: arterial change and vasoconstrictor substances in cerebrospinal fluid. In *Cerebral microcirculation and metabolism* (ed. J. Cervos-Navarro and E. Fritschka), pp. 359–60. Raven Press, New York.

Hughes, J. T. (1984). Regeneration in the human spinal cord: a review of the response to injury of the various constituents of the human spinal cord. *Paraplegia* **22**, 131–7.

3

Radiology of spine and spinal cord injury

P. L. Cook

Introduction

The leading part played by diagnostic radiology in the initial diagnosis and the subsequent management of patients with injuries to the spine and spinal cord is self-evident. Clinical decisions are usually based on a radiological evaluation. This in turn depends upon high-quality radiographs. A plain-film examination in two planes is the minimum requirement. In the examination of neck injuries inclusion of the seventh cervical vertebra is mandatory (Fig. 3.1). The superimposition of soft tissues, particularly the shoulders at

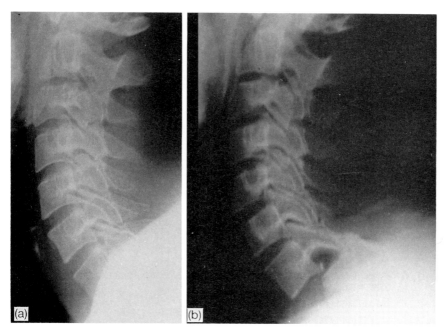

(a)

(b)

Fig. 3.1. (a) A lateral radiograph of the cervical spine does not include the entire seventh cervical vertebra (C7) and shows only a fracture of the C6 facet. (b) A second radiograph made after further traction reveals a bilateral C7–T1 facet dislocation.

41

the cervicodorsal junction, together with the clinical condition and the neurological status of the patient, may make it necessary to supplement these basic views with additional projections or other imaging techniques, after an initial radiological assessment has been made.

Radiography

Tables 3.1 and 3.2 list the radiographic views and techniques that may be necessary for full radiological examination of the cervical and thoracolumbar spine, respectively. Using isocentric techniques these views can be obtained with the patient in the supine position and with the minimum of movement. A fluoroscopic facility is a further aid to positioning for plain films and tomography and allows the smallest possible field to be used, thus improving definition and reducing radiation exposure. With skilled

Table 3.1. Radiographic evaluation of injuries of the cervical spine

---

   (1) Lateral—base of skull through C7
   (2) Antero–posterior—15° of cranial angulation
   (3) Antero–posterior—25–30° of caudal angulation
   (4) Antero–posterior—open-mouth
   (5) Lateral cervicodorsal junction if not shown on (1)
   (6) Supine oblique views (a) for neural arches
                                    (b) for articular pillars
   (7) Lateral views in flexion–extension
   (8) Plain tomography
   (9) Computed tomography
  (10) Myelography ± further computed tomography
  (11) Magnetic resonance imaging

Table 3.2. Radiographic evaluation of injuries of the thoracolumbar spine

---

  (1) Lateral—thoracic and lumbar spine
  (2) Antero–posterior—thoracic and lumbar spine
  (3) Lateral —cervicodorsal junction
               —lumbosacral junction
  (4) Antero–posterior lumbosacral junction (10–25° of cranial angulation)
  (5) Antero–posterior and lateral sacrum
  (6) Supine oblique views
  (7) Plain tomography
  (8) Computed tomography
  (9) Myeloradiculography ± further computed tomography
 (10) Magnetic resonance imaging

---

assistance it is possible to move a spinally injured patient in complete safety; for example, when radiography is being undertaken with mobile apparatus. It is helpful under these circumstances to use a non-grid technique, thus enabling the use of a finer focus and shorter exposure times and allowing the cassette to be placed at an angle other than normal to the X-ray beam (Woodford and Walford 1985). For subsequent examinations of paraplegic or tetraplegic patients the use of specially designed mattresses is recommended to restrict skin pressures to below 50 mm Hg and to lessen the risk of pressure sores compared with standard radiographic mattresses.

The multiplicity of recommended views of the cervical spine, further to analyse bony injury, is an acknowledgement that dependence on the lateral projection will result in errors of diagnosis (Gehweiler, Osborne, and Becker 1980), but pre-dates the general use of computed tomography whenever significant bony or ligamentous injuries are shown on the plain-film examination. Penning (1968), whilst not condoning the restriction of plain-film assessment to this degree, nevertheless regarded the lateral projection as paramount. He emphasized that the lateral radiograph permits the diagnosis of all major traumatic lesions of the cervical spine and therefore provides the best immediate information. Gehweiler *et al.* (1980) estimate that two-thirds of cervical spine fractures and dislocations can be recognized on the lateral radiograph, but emphasize that injuries with similar appearances may require very different management (Fig. 3.2).

Plain tomography may be rectilinear or polydirectional. The use of an apparatus enabling coronal or sagittal tomography without moving the patient, together with high-speed film–screen combinations allowing the use of fine focus and low kilovoltage, is more important than the production of ultra-thin sections. The latter may be difficult to orientate, and zonographic

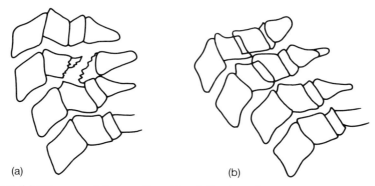

(a) (b)

Fig. 3.2. (a) Diagram of the recoil position following a hyperextension injury of the cervical spine. (b) Diagram showing an apparently similar position due to unilateral facet dislocation.

(6° thick-section) techniques are preferred at the Wessex Spinal Injuries Unit (Woodford and Walford 1985). The role of plain tomography is decreasing as the improved resolution of reformatted coronal and sagittal images on computed tomography eliminates the major diagnostic limitations of sections made only in the axial plane.

Patterns of injury

Injuries of the spine and spinal cord both show peaks of incidence at the low cervical and thoracolumbar levels (Jefferson 1927; Rogers 1981). For spinal injuries each peak comprises about 30 per cent of the total with a third, smaller peak at the atlanto–axial level that is not associated with any corresponding increase in damage to the cervical cord in surviving patients, although there is a considerable variation in different series. For spinal cord injuries there is an even greater predominance of injuries at lower cervical levels with approximately 40 per cent occurring between the fourth and seventh cervical vertebrae. A further 30 per cent of cord injuries occur between the tenth thoracic and second lumbar levels. Spine and cord injuries are otherwise almost uniformly distributed segment by segment.

Injuries sustained in different ways result in major differences in the level at which the cord is damaged. Pedestrians, bicyclists, motor cyclists, and particularly the occupants of motor vehicles show a predilection for cord injuries at the level of the atlas. Those sustained in jumps or falls are seldom seen at this level and those from diving accidents almost never. Cord injuries are typically at lower cervical levels when resulting from these latter types of accident (Fife and Kraus 1986). Calenoff, Chessare, Rogers, Toerge, and Rosen (1978) found bony injuries at more than one spinal segmental level in 16 per cent of patients and these were discontiguous in 4 per cent. The primary fractures occur with disproportionate frequency in the upper and mid-dorsal spine. The secondary injuries are often found at the atlanto–axial or lumbosacral levels, i.e. at the extremities of the spine, at which levels they may readily escape radiological evaluation. They are particularly likely to occur in patients with severe neurological deficits. In patients who are paraplegic from an upper or mid-dorsal fracture–dislocation a second vertebral injury is present in two-thirds of patients. Similarly, Fife and Kraus (1986) in a series of 550 patients identified a second cord injury, separated by at least one uninjured spinal segment from the highest cord lesion, in 28 per cent of patients.

Excluding vertebral collapse due to senile, post-menopausal, or steroid-induced osteoporosis or metastatic destruction, about 10 per cent of patients with spinal injuries show evidence of damage to the spinal cord. Conversely, about 10 per cent of patients with cord damage show no radiographic evidence of a spinal fracture or dislocation (Riggins and Kraus 1977; Rogers

Table 3.3. Correlation of injuries to the spine and spinal cord

Level of injury	Cord damage (per cent)	Neurologically complete (per cent)
Cervical	40	10
Thoracolumbar		
(a) Wedge-compression	5–10	—
(b) Fracture–dislocation	60–75	50–60

1981). In the latter group the effects of minimal trauma on pre-existing spondylosis or a constitutionally narrow spinal canal are particularly significant (Hardy 1977; Epstein, Epstein, Benjamin, and Ransohoff 1980; Eismont, Clifford, Goldberg, and Green 1984). Table 3.3 summarizes the correlations between spine and spinal cord injuries from several series. Approximately 40 per cent of patients with injuries of the cervical spine show evidence of damage to the cord (Rogers 1981) and of these rather more than one-quarter are neurologically complete. Only about 10 per cent of injuries to the dorsal spine are associated with a cord injury, but when these occur they are more likely to be severe: more than three-quarters are completed neurologically. Most injuries at the thoracolumbar junction are simple wedge-compression fractures with only a small risk of injury to the cord or *conus*. Patients with unstable or malaligned fractures or dislocations have a much higher risk of cord damage, rising to 70 per cent in some series (Kaufer and Hayes 1966; Burke 1971b; Harris 1978a). Many of these are major cord injuries.

Spinal stability

The biomechanical and clinical concepts of spinal stability after injury remain confusing and are equally difficult to assess radiologically. Nicoll (1949) and Holdsworth (1963), from studies of thoracolumbar injuries, mainly sustained in mining accidents, laid down the foundations of subsequent perceptions of the spine as a two-column structure by defining fractures limited to the vertebral body as stable, but those in which there was additional damage to the posterior ligamentous complex or the bony elements of the neural arch as unstable. These elements of instability were seldom, if ever, caused by pure flexion or compression violence, but frequently resulted when the spine was forcibly rotated. There was then a potential for initiating cord damage or increasing the severity of a cord injury if one were already present. Holdsworth (1963) drew frequent parallels with cervical injuries and obviously regarded stability of the cervical spine as dependent on similar factors.

Subsequently, Kelly and Whitesides (1968) regarded the thoracolumbar spine as a two-column structure with a solid anterior column comprised of the vertebral bodies and a hollow posterior column made up of the neural arches. Distinctions between anterior and posterior instability were made and regarded as relevant to surgical management. The two-column concept, with some modifications, has been more formally extended to include the cervical spine (White, Johnson, Panjabi, and Southwick 1975; White, Southwick, and Panjabi 1976). In 1976 Webb, Broughton, McSweeney, and Park stated that the following X-ray signs, either singly or in combination, may indicate instability of the cervical spine. These are: interspinous widening; vertebral subluxation; vertebral compression fracture; loss of cervical lordosis.

White and Panjabi (1978) identified more specific features and defined radiographic instability as horizontal displacement greater than 3.5 mm at any segmental level on the lateral film or an angular difference 11° greater than that at adjacent cervical segments. White and Panjabi (1978) also attempted to quantify thoracolumbar instability by the use of a scored check-list (Table 3.4), but this lacks clinical verification. In 1977 Louis considered that the posterior column comprised the pair of articular pillars, but Panjabi, Hausfield, and White (1981) reiterated the importance of ligamentous injuries, the radiographic features of which were further stressed by Green, Harle, and Harris (1981).

The definition of instability has been further widened to include conditions threatening, or giving rise to, neurological deficit, spinal deformity under physiological loads, or mechanical neck and back pain at a later date—delayed instability (Cheshire 1969; Roberts and Curtis 1970; McAfee, Yuan, and Lasda 1982). These may follow purely discoligamentous injuries (Cheshire 1969) or burst fractures. The latter have been regarded as stable by Holdsworth (1963) as they were thought not to be associated with posterior ligamentous disruption. Not infrequently, however, these frac-

Table 3.4. Instability of thoracolumbar injuries (a score >5 indicates instability) (White and Panjabi 1978)

Injury	Score
Anterior element disruption	2
Posterior element disruption	2
Sagittal translation >2.5 mm	2
Rotation >5°	2
Damage to spinal cord or *cauda equina*	2
Costovertebral disruption	1
Anticipated dangerous loading	1

Two-column

Ant.   Post.

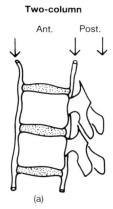

(a)

Three-column

Ant. Mid. Post.

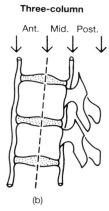

(b)

Fig. 3.3. The two-column and three-column concepts of spinal stability (after Denis 1983). (a) The vertebral bodies, intervertebral discs, and longitudinal ligaments comprise the anterior column. The posterior column is made up of the ligamentous complexes and bony arches. (b) The anterior column is separated into an anterior and middle column.

tures are found to have complete ligamentous tears (Whitesides 1977), or unsuspected fractures of the neural arch are shown on computed tomography (McAfee *et al.* 1982). Denis (1983) considered that there had been a failure to appreciate instability resulting from tears of the posterior longitudinal ligament and the posterior segments of the annulus. The instability of distraction injuries and burst fractures had consequently been underestimated. Using the additional information provided by computed tomography on a series of more than 400 thoracolumbar injuries, Denis (1983) therefore proposed that the two-column theory should be replaced by a three-column concept. This is outlined in Fig. 3.3. Although originally directed towards the thoracolumbar spine, Harris, Edeiken-Monroe, and Kopaniky (1986) indicate the relevance of this revision to cervical spine injuries. The basic principle of the two-column theory is unaltered, namely that, during flexion, anterior compression forces are accompanied by posterior distraction. Thus the mechanistic approach has not been abandoned, but the separation of the anterior column of Holdsworth (1963) into an anterior and middle column contributes to a better understanding of distraction fractures as unstable, with disruption of the middle and posterior columns. It similarly facilitates the differentiation of burst fractures into stable or unstable types, according to whether disruption of the middle column accompanies the anterior column compression. The assessment of stability according to this scheme cannot be made satisfactorily without computed tomography. By this means Denis (1983) regards fractures of the vertebral appendages, and minimal to moderate compression fractures of the vertebral bodies, as stable. The grading of instability is shown in Table 3.5. These have helped to distinguish between acute and late mechanical and neurological instability. Nevertheless, evaluation of fracture patterns in individual patients may be problematical and other assessments continue to be made (Kingma 1986) (see Table 3.6). Beyond considerations of stability the classification of injuries to the cervical and thoracolumbar spine is based on the mechanism by which they are produced. The present functional classification of cervical spine injuries is based on the work of Whitley and Forsyth (1960). Similarly, the extension by Holdsworth (1963) of Nicoll's (1949) classification of thoracolumbar injuries has required little modification with the passage of time and has the merit of simplicity.

Table 3.5. Instability of thoracolumbar injuries (Denis 1983)

Injury	Attached risk
First degree—mechanical instability	Development of spinal deformity
Second degree—neurological instability	Development of neurological deficit
Third degree—mechanical and neurological instability	Worsening of spinal deformity or of neurological deficit already present

Table 3.6. Instability of thoracolumbar injuries (Kingma 1986)

Grades of injury	Injury
0	Isolated fractures of the spinous or transverse processes Isolated anterior chip fractures without dislocation
1	Anterior wedge fractures of less than 15°
2	Isolated wedge fractures of more than 15° without discoligamentous injury Isolated lateral wedge fractures of less than 15°
3	Isolated wedge fractures with disruption Multiple wedge fractures Lateral wedge fractures of more than 15° Injuries involving the posterior part of the vertebral body
4	Bony and/or ligamentous injuries of the anterior and posterior vertebral columns, including all subluxations
5	Serious disruption of the anterior and posterior vertebral columns, including all dislocations

Cervical spine injuries

Complex classifications of these injuries have been formulated, in attempts to include all possible mechanisms, but, most simply, the predominant forces are those of hyperflexion, hyperextension, axial compression, and rotation with or without the superimposition of a shearing stress. In direct hyperflexion the energy of mild forces is dissipated when the chin strikes the sternum. At this point the movement of the neck is arrested before major bony or ligamentous injuries are sustained. With greater forces, however, the impact of the chin creates a new fulcrum on the anterior chest wall. The mechanisms are then analogous to seat-belt injuries of the lumbar spine, with greatly increased forces of posterior distraction and damage to the posterior ligamentous complexes (Roaf 1960). Furthermore, it is unusual for hyperflexion forces to act in isolation. Rotational forces will result when the head is turned at the moment of injury and there is frequently an additional force applied through a blow to the top, side, or back of the head. This will impose, respectively, an axial, lateral flexional/rotational, or rotational/shearing stress on the hyperflexed cervical spine. The bio-mechanical experiments of Roaf (1960) have established that the longitudinal ligaments and the posterior ligamentous complex will not tear without these additional stresses, which are therefore of particular significance in the production of an unstable injury. Elaborate analyses of the dynamics of spinal injuries (Roaf 1972) identify the factors that operate in the production of injuries of different types. Their complexity limits their usefulness in clinical practice. Paradoxically, the complexity of movements of the head and neck in the first few milliseconds after an impact may almost defy analysis, when such unknown and extraneous factors as the resilience of a car seat may be relevant (McKenzie and Williams 1971). The importance of hyperextension injuries of the neck, which make up about 30 per cent of the total, was recognized in the mid-nineteenth century (Malgaigne 1855), from clinical and anatomical observations, but was thereafter overlooked until redescribed by Taylor and Blackwood (1948) and Barnes (1948). The presence of cord damage in the absence of radiographic evidence of a bony injury was noted by these authors in 17 per cent of patients in the series of Riggins and Kraus (1977). This initially led to suggestions that compression was due to compromise of the spinal canal by disc protrusion or by corrugation of the ligamenta flava. It has since been appreciated that the presence of a small spinal canal may greatly influence the outcome and also that the positions reached at the moment of injury, in the unstable spine, bear little relation to the 'recoil position' at the time of radiographic examination (Swain, Grundy, and Russell 1986) (Figs. 3.4 and 3.5).

The major injuries of the cervical spine at the craniovertebral and lower cervical levels are listed in Tables 3.7 and 3.8 (Gehweiler *et al.* 1980; Harris

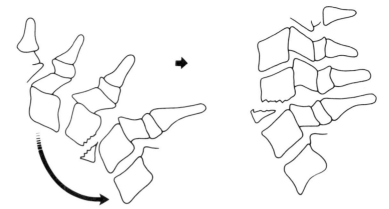

Fig. 3.4. Diagram of hyperflexion cervical spine injury showing posterior alignment of upper vertebrae in the recoil position.

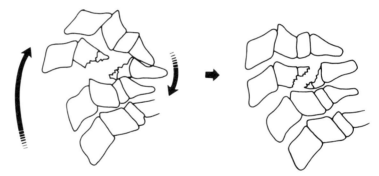

Fig. 3.5. Diagram of hyperextension cervical spine injury showing anterior alignment of upper vertebrae in the recoil position.

et al. 1986). Gehweiler, Clark, Schaaf, Powers, and Miller (1979) and Gehweiler *et al.* (1980) have produced simplified groupings of the hyperflexion and hyperextension injuries (Figs. 3.6 and 3.7).

In addition to the direct evidence of fracture or dislocation of the vertebral bodies and neural arches, a number of indirect radiographic signs of cervical spine injury have been described (Weir 1975; Clark, Gehweiler, and Laib 1979). These may provide a clue to underlying trauma and its mechanism and fall into two main categories: (1) abnormalities of the pre-vertebral soft tissues due to haematoma; (2) abnormalities of the relationships and alignment of the vertebrae due to discoligamentous injuries. Malalignment may be seen in the antero–posterior or lateral projections and may affect the vertebral bodies, the posterior joints and the articular pillars, or neural

Table 3.7. Injuries of the upper cervical spine

Jefferson fracture
Hangman fracture
Atlanto-occipital and atlanto–axial dislocation ⎤ anterior or
Odontoid fracture and atlanto–axial fracture dislocation ⎦ posterior

Table 3.8. Injuries of the lower cervical spine

Wedge fracture
Burst fracture
Clay shoveller's fracture
Fracture of articular pillar
Momentary dislocation (capsular sprain) ⎤ hyperflexion or
Fracture dislocation ⎦ hypertension
Facet dislocation—unilateral or bilateral

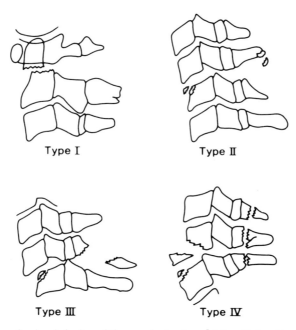

Fig. 3.6. Hyperflexion injuries of the cervical spine. [After Gehweiler *et al.* (1979).] Type I: Odontoid fracture with anterior displacement. Type II: Momentary dislocation with predominantly ligamentous injury. Type III: Bilateral dislocation of the facet joints. Type IV: Hyperflexion 'tear-drop' fracture–dislocation.

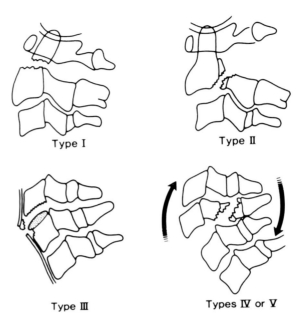

Type I   Type II

Type III   Types IV or V

Fig. 3.7. Hyperextension injuries of the cervical spine. [After Gehweiler *et al.* (1979).] Type I: Odontoid fracture with posterior displacement. Type II: Hangman fracture. Type III: Momentary dislocation with disruption of the discovertebral bond and predominantly ligamentous injury. Types IV and V: Increasing ligamentous injury with compression and comminution of the articular pillar and posterior bony arch.

Table 3.9. Radiographic signs of instability in cervical spine injuries

Disruption of intervertebral disc
Disruption of an apophyseal joint or a facet fracture
Separation of laminae or spinous processes on lateral radiograph
Malalignment of pedicles or spinous processes on frontal radiograph

arches. Some types of injury are inherently unstable but others may vary. The radiographs should be examined for evidence of instability (Table 3.9) in every cervical injury no matter how minor.

Atlanto–axial injuries

About a third of cervical spine injuries occur at this level and all the major injuries are unstable. Injuries of the axis occur approximately four-times as frequently as those of the atlas.

Jefferson fracture

Described by Jefferson in 1920, this is a bursting injury of the atlas resulting from axial compression between the occipital condyles and the lateral atlanto–axial articulations (see Figs. 3.8 and 3.9). The lateral masses of the atlas are forced apart and the anterior and posterior arches fracture, typically on both sides of the midline. The transverse ligament is usually torn, either centrally or with avulsion of a small bone fragment at its lateral attachment. With minor injuries the ligament will stretch but remain intact and the disruption does not involve the capsular ligaments of the atlanto–axial facet joints. Gross instability is prevented although the fractured atlas no longer forms a fully stable ring around the odontoid. With more severe injuries the capsular ligaments are also torn and displacement and instability are more marked. The alar ligaments remain intact but are relatively weak and cannot maintain stability (Fielding, von Cochran, Lawsing, and Hohl 1974).

Radiographic projections through the open mouth, or tomograms, show that the distance between the odontoid and the displaced lateral masses is greater than the normal 3 mm on each side and that the normal lateral alignment between the atlas and the axis is lost (Fig. 3.9). If the latter displacement totals more than 7 mm on the two sides, the transverse ligament is likely to be torn (Spence, Decker, and Sell 1970). An axial force applied to the tilted head may result in a unilateral fracture with displacement of only one lateral mass. The odontoid relationships may then be more

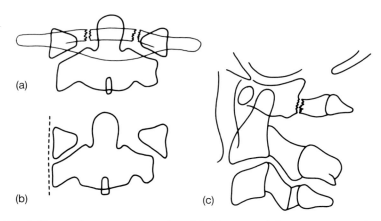

(a)

(b)

(c)

Fig. 3.8. Jefferson fracture of the atlas. (a) Diagram of the antero–posterior view showing bilateral fractures of the anterior arch and displacement of the lateral masses. (b) Composite diagram of atlanto–axial relationships—normal (broken line) and following a burst fracture (right). (c) Diagram of lateral view showing a fracture of the posterior arch.

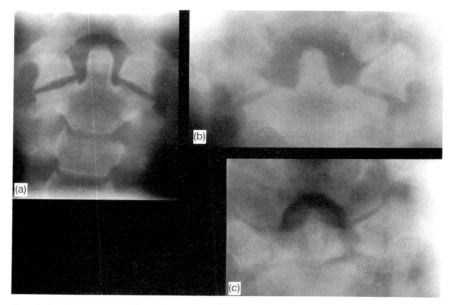

Fig. 3.9. Antero–posterior atlanto–axial tomograms. (a) Normal, for comparison. (b) Jefferson fracture showing lateral displacement in relation to the axis and odontoid. (c) Jefferson fracture showing fractures of the anterior arch.

difficult to differentiate from the asymmetry resulting from rotation of the head or from rotary atlanto–axial fixation or subluxation (Fielding and Hawkins 1977). In these, however, the lateral alignment of the atlas and axis should be maintained. This lateral relationship may briefly be disturbed in childhood, following a period of differential growth of the atlas and axis, but this is usually of a minor degree and should not be a diagnostic problem. With rotary fixation, which in childhood is self-limiting and is more likely to follow a viral illness than trauma, the head and neck are held tilted, rotated, and slightly flexed—the 'Cock-Robin' position. Radiographically the open-mouth view shows asymmetry of the lateral atlanto–axial joints due to subluxation and overriding on one side (Levine and Edwards 1986).

More than 50 per cent of patients with fractures of the atlas, even when these involve only the posterior arch, show evidence of another cervical fracture. Most commonly these are hangman fractures or odontoid fractures of the hyperextension type (Fig. 3.10). Neurological signs and symptoms are more likely to result from these injuries than from the atlas fracture itself (Levine and Edwards 1986). Fractures of the anterior or posterior arch, particularly the latter, are not usually difficult to identify on plain radio-graphs but, if there be any doubt, the vertically orientated fractures are particularly well shown on computed tomographic scanning in the axial

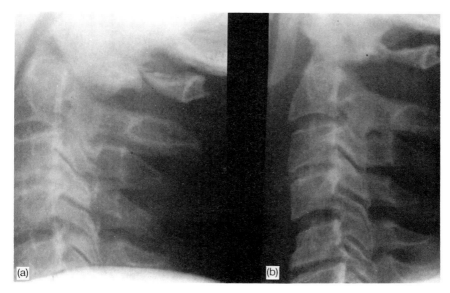

(a) (b)

Fig. 3.10 Lateral radiographs showing a fracture of the posterior arch of the atlas sustained in a fall downstairs. The patient showed a motor deficit and there is an associated hyperextension fracture of the odontoid with realignment in the flexed position (b).

plane. Computed tomography may also facilitate the differentiation of fractures from congenital clefts and defects in the arches of the atlas.

Hangman fracture

This type of fracture has been described by Haughton (1866) and Wood-Jones (1913) (see Figs. 3.11, 3.12, and 3.13). The axis is the only cervical vertebra in which the superior and inferior articulations are not vertically in line. The short pedicle forms a *pars interarticularis* and the hangman fracture is a bilateral traumatic spondylolysis. The fracture separates the head, atlas, odontoid, and body of the axis from the C2–3 facet joints, and they are therefore not stabilized, through the articular pillars to the lower cervical spine, and are free to displace anteriorly. Separation through the fracture results in a traumatic spondylolisthesis (Schneider, Livingston, Cave, and Hamilton 1965).

Hangman fractures have been classified into three types (Levine and Edwards 1986) according to the angulation and the extent of translation and separation through the fracture (Table 3.10). This grading is relevant to decisions regarding adoption of conservative or surgical management as it correlates well with the differing mechanisms of injury and the frequency of a neurological deficit. The more severe injuries may have been accompanied

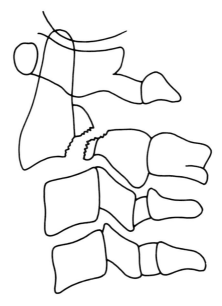

Fig. 3.11. Diagram of hangman fracture of the axis.

by an element of flexion with wide separation as a traumatic spondylolis-thesis. This may also be associated with disruption of the C2–3 disc and the posterior longitudinal ligament, ultimately with unilateral or bilateral dislo-cation of the C2–3 facet joints. The injury then more closely resembles that of judicial hanging (Grogono 1954) and not surprisingly shows a much higher incidence of cord damage. Immediate surgical reduction and stabilization is necessary, whereas the minimally displaced and angled fractures can be shown with flexion–extension views, made under medical supervision, to be almost stable. These fractures will invariably unite spontaneously without complication. The fracture line is easily seen on the lateral radiograph [Fig. 3.13(a)]. It is also well demonstrated [Figs. 3.13(b) and (c)] on computed tomography, which may also show any fractures involving the facet joints. The presence of any upper cervical fracture of the hyperextension type should suggest the possibility of a concomitant hang-man fracture. Such injuries include the avulsion of a small fragment from the base of the body of the axis by the anterior longitudinal ligament—the so-called hyperextension tear-drop injury, and the fractures of the posterior arch of the atlas or of the spinous process of the axis.

Neither the Jefferson nor the hangman fracture characteristically gives rise to a severe or permanent neurological deficit, although the frequency of neurological complications in the latter has varied from 6.5 to 73 per cent in

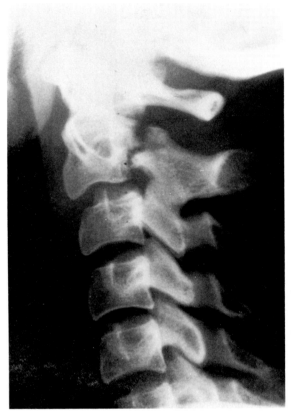

Fig. 3.12. Lateral radiograph of a moderately separated hangman fracture (type II).

different series (Schneider *et al.* 1965). The symptoms are predominantly local—neck and suboccipital pain. Turning the head may produce pain or tingling in the fingers, but the presence of a more severe deficit is more likely to be due to a concomitant injury at another level. Difficulty in swallowing or breathing is usually due to a retropharyngeal haematoma rather than to compression of the brain stem or cervicomedullary junction, but it has been suggested that distortion and stretching of the vertebral arteries may produce intimal tears with subsequent spasm or thrombosis. This can cause vertebrobasilar ischaemia and lower cranial nerve palsies (Grundy, McSweeney, and Jones 1984; Pelker and Dorfman 1986), or transient symptoms and signs such as rotatory nystagmus, delayed diplopia, or visual loss. The infrequency of neurological complications is probably due to the wide spinal canal at upper cervical levels relative to the cord, which occupies

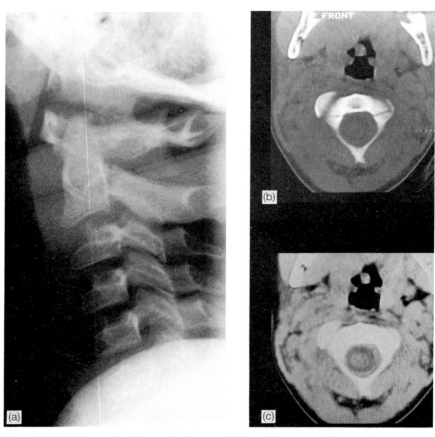

Fig. 3.13. (a) Lateral radiograph of an undisplaced hangman fracture (type I) in a five-year-old boy who slipped under an ill-fitting seat belt in a road traffic accident. (b) and (c) Computed tomographic scans of the same injury at different window settings showing the fracture (b) and the minimal displacement of the dural tube (c).

Table 3.10. Radiographic classification of hangman fractures (Levine and Edwards 1986)

Type	Description
I	Undisplaced or showing no angulation and less than 3 mm displacement
II	More than 3 mm displacement and angulation
IIa	Less displaced but more severely angulated
III	Severe displacement and angulation with disruption of the C2–3 disc and unilateral or bilateral dislocation of the C2–3 facets

only about 25 per cent of the cross-sectional area. Any separation of the Jefferson or hangman fracture further widens the canal.

Odontoid fractures and fracture–dislocations

These may be sustained in flexion or extension (Fig. 3.14). The latter is slightly more common and more likely to be associated with other cervical injuries. The injuries are unstable as the lateral atlanto–axial joints do not provide any significant stability, either in flexion or extension. They may indeed already be subluxed to the extent that there is any displacement or angulation of the odontoid and atlanto–axial malalignment. Such displacement reduces the sagittal diameter of the spinal canal. Neurological symptoms and signs are therefore somewhat more common than with Jefferson or hangman fractures. They occur more frequently with anteriorly displaced fractures (Paradis and Jones 1973) resulting from hyperflexion. The upper cord, or the C1–2 nerve roots, may be directly compressed or the deficit may result from compression of the anterior spinal arteries. Nevertheless, local symptoms, including torticollis, again predominate. Under the age of seven years the fracture may pass through the subdental synchondrosis as a traumatic epiphysiolysis. This may result in resorption of the odontoid.

The presence of an upper cervical injury may be overlooked when the clinical picture is dominated by additional injuries to the skull, face, or chest. The diagnosis may then be delayed or the injury discovered incidentally. A lateral radiograph of the neck should be made in all patients who have suffered a head injury severe enough to cause a period of unconsciousness (Evans 1976) or who will require a general anaesthetic for the manage-

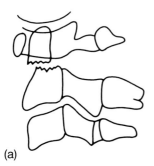

(a)   (b)

Fig. 3.14. Fracture of the odontoid. (a) Lateral diagram of anteriorly displaced fracture. The spinal canal is narrowed by the forward displacement of the atlas, particularly if the transverse ligament is torn. (b) Lateral diagram of posteriorly displaced fracture. The spinal canal is narrowed between the odontoid and the posterior arch of the axis but the transverse ligament is less likely to be torn in hyperextension.

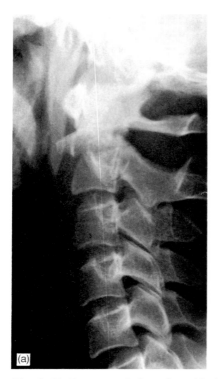 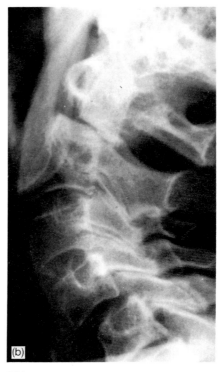

Fig. 3.15. Fracture of the odontoid. (a) This young man attended the accident department complaining of neck pain several days after a road traffic accident. The lateral radiograph shows an anteriorly displaced odontoid fracture. (b) The more displaced posterior odontoid fracture shown in this lateral radiograph was seen initially on the lateral skull radiograph following a head injury. Neither patient showed a neurological deficit.

ment of other injuries. Only by these means can the occult injuries to the upper cervical spine be identified before neurological complications have developed (Paradis and Jones 1973; Chakera, Anderson, and Edis 1982). The diagnosis may be further delayed by the paucity of neurological signs. In one series (Nachemson 1960) one-third of patients with odontoid fractures did not consult a doctor for several days after the injury (Fig. 3.15).

Fractures of the odontoid have been classified into three types (Anderson and d'Alonzo 1974) (Table 3.11). The type-I injury is rare, and type-II and type-III injuries occur with approximately equal frequency. An undisplaced type-III fracture is virtually stable and almost invariably unites, although posterior angulation may give rise to late *sequelae*. There is a considerable risk of non-union in type-II fractures, with no statistically significant difference between those displaced anteriorly or posteriorly. Union rates in

Table 3.11. Radiographic classification of odontoid fractures (Anderson and d'Alonzo 1974)

Type	Description
I	Avulsion of a small fragment at the site of attachment of the alar ligaments
II	Through the base of the axis with anterior or posterior displacement and angulation
III	Through the cancellous bone of the body of the axis

non-surgically treated patients range from 5 to 64 per cent (Schatzker, Rorabeck, and Waddell 1971). Earlier reports (Roberts and Wickstrom 1973; Apuzzo, Heiden, and Weiss 1978) suggested that non-union would result when there was a 5 mm or more displacement through the fracture. More recently a significant increase in non-union has been noted with a 2 mm or more displacement (Wang, Mabie, Whitehill, and Stamp 1984). Non-union of odontoid fractures may result in persistent or delayed instability. Angulation and displacement may increase progressively or follow minor trauma (Fig. 3.16). The malalignment narrows the upper

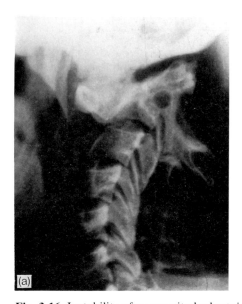

Fig. 3.16. Instability of an ununited odontoid fracture. The odontoid fracture in this patient was diagnosed only in retrospect, a year after a neck injury, when increasing deformity and the development of a cervical myelopathy were followed by this lateral radiograph. The marked angulation and displacement through the ununited fracture narrows the spinal canal. Lateral radiograph (a) and tomogram (b).

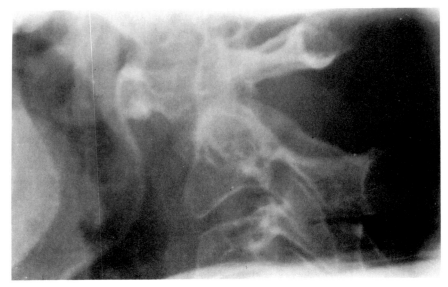

Fig. 3.17. Atlanto–axial instability in rheumatoid arthritis. This patient showed increased deformity of the neck and signs of upper cervical cord compression following a whip-lash injury. The lateral radiograph shows marked atlanto–axial subluxation with no bony injury.

cervical canal and compresses the cervicomedullary junction and upper cervical cord, causing a delayed myelopathy with the development of long-tract neurological signs. The clinical picture is similar to that resulting from atlanto–axial subluxation from other causes such as rheumatoid arthritis or sero-negative spondylarthropathies, paediatric infections, Down's syndrome, or mucopolysaccharidoses. In all such patients trauma may aggravate the pre-existing condition (Fig. 3.17). Type-II odontoid fractures may be difficult to differentiate clinically or radiologically from an *os odontoideum*. This is a developmental anomaly in which the ossification centre for the odontoid fails to fuse with the base of the axis. The condition results in instability, but trauma is often an additional triggering factor (Fig. 3.18).

In spite of the mild symptomatology and delayed diagnosis in some patients, upper cervical fractures can result in a devastating cord injury (Fig. 3.19). Approximately 30 per cent of the fatally injured victims of road traffic accidents show injuries at the craniovertebral level (Alker, Oh, Leslie, Lehotay, Panaro, and Eschner 1975). Nevertheless, the survivors seldom show a severe or permanent neurological deficit. When the latter occurs, following any of the above injuries, it usually takes the form of a motor paresis without sensory loss. Some patients show alterations in mental or

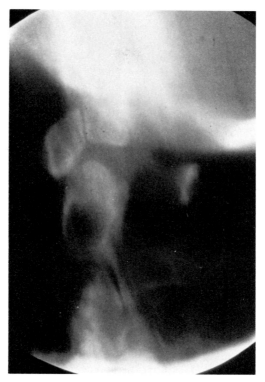

Fig. 3.18. *Os odontoideum.* This patient was thrown through the roof of a car in 1959 and was transiently tetraplegic. An odontoid fracture was diagnosed at another hospital and was thought to be stable following three months' immobilization. In 1985 he started to show signs of upper cervical cord compression and the previously diagnosed odontoid fracture was shown on this lateral tomogram to be an *os odontoideum.*

motor behaviour and there is occasionally change in the pattern of respiration with a loss of voluntary spontaneous control.

Lower cervical injuries

Introduction

Injuries of the lower cervical spine are listed in Table 3.8. In contrast to craniovertebral injuries, most lower cervical injuries are inherently stable. A detailed analysis, including computed tomography, of 400 cervical injuries (Miller, Gehweiler, Martinez, Charlton, and Daffner 1978) revealed that assessments from plain radiographs underestimate the frequency and extent of injuries to the posterior bony arch compared with

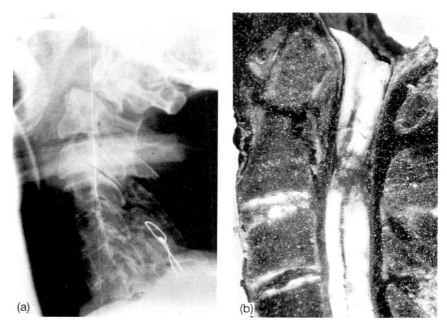

Fig. 3.19. Odontoid fracture and major cord injury. This 60-year-old riding instruc-
tress was kicked in the face by a horse. She was immediately tetraplegic and died
shortly after reaching hospital. (a) Lateral radiograph showing posteriorly displaced
odontoid fracture (she had previously fractured her lower cervical spine in another
riding accident). (b) Post-mortem specimen. [Taken with permission from Weller,
Swash, McLellan, and Scholtz (1983).]

the vèrtebral body. In this series 50 per cent of patients showed fractures of
the posterior elements, which, in common with those of the body, most
frequently affected the fifth and sixth cervical segments. In many normal
patients the bodies of C4 and C5 are slightly flatter than those above and
below. This is not usually associated with wedging, but nevertheless may be
interpreted as vertebral compression. The presence of a wedge fracture
should not be inferred unless the anterior height of the vertebral body is at
least 3 mm less than the posterior height. In spondylotic patients, bone
accretion anteriorly may also result in a spurious impression of vertebral
collapse.

Ligamentous injuries
Momentary dislocations or sprains of the capsular ligaments of the
apophyseal joints (Braakman and Penning 1971; Hardy 1977; Epstein *et al.*
1980) may be sustained in hyperflexion or hyperextension. These may show
only minimal radiographic signs of injury when examined in the recoil

position and may therefore appear to be stable. The secondary radiographic signs of soft tissue and ligamentous injury may then be particularly valuable in the acute phase. Some disruption of the posterior ligamentous complex is an integral part of all hyperflexion injuries. This may be seen radiographically as separation through the apophyseal joints, or of the laminae and spinous processes. In hyperextension the spinous process ultimately acts as a fulcrum, with stretching or rupture of the anterior longitudinal ligament and disruption of the intervertebral discs. Radiographically this may be revealed as pre-vertebral soft tissue swelling or occasionally as a gas-vacuum phenomenon. Again, there is frequently no evidence of bony injury, but in some patients a small fragment may be avulsed from the superior corner of the vertebral body anteriorly. These hyperextension sprains may result in anterior, central, or complete cord syndromes (Fig. 3.20) as a result of cord compression within the traumatic pincers described by Taylor and Blackwood (1948). In this, the spinal canal is compromised anteriorly by horizon-

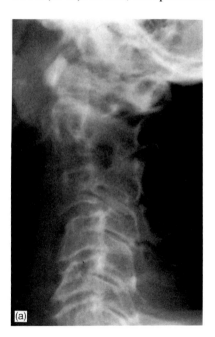
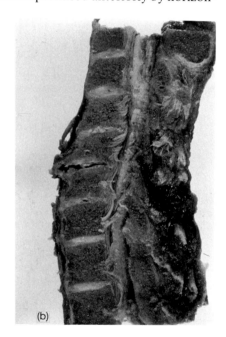

Fig. 3.20. Hyperextension injury of the cervical spine. This 72-year-old woman was immediately tetraplegic following a fall downstairs. The radiograph (a) shows disc degenerative and spondylotic changes, but the only evidence of injury was pre-vertebral soft tissue swelling consistent with hyperextension damage to the anterior longitudinal ligament. The pathological specimen (b) shows extensive ligamentous damage and disruption of the C5–6 disc typical of hyperextension with a total cord injury. [Taken with permission from Weller (1984).]

tal translation of the vertebrae and extrusion of disc fragments, and posteriorly by corrugation of the ligamenta flava and malalignment of the laminae. The cord is more likely to be compressed when the spinal canal is small or when spondylotic changes are present (Fig. 3.20).

Delayed instability following capsular sprains is found in up to 22 per cent of patients (Cheshire 1969) as a reflection of the poor healing capacity of ligaments (Holdsworth 1963). This delayed instability may give rise to a myelopathy, but more often results in local suboccipital, neck, interscapular, or radicular pain, paraesthesia, or numbness—the whip-lash syndrome (Table 3.12). Many of the clinical features of this condition are subjective and inconstant, so that their validity has been questioned and they have been regarded as a manifestation of compensation neurosis (Miller 1961). Such doubts have been rebutted clinically and experimentally (MacNab 1971; Hohl 1974; Merskey 1984) (Table 3.13). The head is held rigidly with some reversal of the normal cervical lordosis on the lateral radiograph—the 'West Point' position. This posture, together with restricted motion at one interspace, is regarded as an adverse prognostic feature by Hohl (1974). White and Panjabi (1978) defined normal ranges of horizontal displacement and angulation between adjacent motion segments, during flexion–extension, as specific indicators of instability, as previously described. These measurements are difficult to evaluate in individual patients

Table 3.12. Clinical features of the whip-lash syndrome

Neck and radicular pain and paraesthesiae
Suboccipital and interscapular pain
Headache
Dysphagia
Tinnitus, buzzing, and popping in the ears
Blurring of vision
Dizziness, vertigo
Impaired memory, lethargy, and fatigue
Objective sensory impairment and reflex changes

Table 3.13. Pathogenesis of the whip-lash syndrome (MacNab 1971)

Spasm or tears of cervical muscles
Tears of anterior longitudinal ligament and discovertebral bond
Damage to posterior ligamentous complex
Retropharyngeal haematoma
Damage to cervical sympathetic and lower cranial nerves

due to overlap with normal measurement (Scher 1979). In a five-year follow-up of 146 patients, Hohl (1974) observed the appearance of disc degeneration changes at the affected level in 39 per cent of the patients, although these changes did not always correlate with the persistence of symptoms. Fifty-seven per cent of patients became symptom-free during the follow-up, but recovery was less likely in older patients. In my department, cineradiography in the erect position has been preferred to flexion–extension radiographs (Brunton, Wilkinson, Wise, and Simons 1982). Abnormal patterns of neck movement have been defined and provide a guide to subsequent management by cervical fusion.

In many normal children the weight of the head and the immature neck musculature produces 'pseudo-subluxation' between the second and third cervical segments and to a lesser extent at segments below this. Sullivan, Bruwer, and Harris (1958) noted a forward glide of 4 mm into flexion between the second and third cervical vertebrae in 9 per cent of normal children, and similar findings have been reported by Cattell and Filtzer (1965). This is frequently misdiagnosed as traumatic subluxation. True subluxation in childhood is usually associated with pre-vertebral soft tissue swelling.

Wedge and burst fractures

Greater forces of hyperflexion will result in a wedge fracture (Fig. 3.21), with impaction and angulation of the superior end-plate and the anterior cortex of the vertebral body. These are normally stable injuries if stability is defined as the potential for further or progressive deformity under physiological loads or during healing, or the threat of progressive encroachment on the spinal canal and injury to the neural structures. However, there may be instability, similar to that of a capsular sprain, although radiographic evidence of stretching or disruption of the ligamentous complexes is easier to identify (Evans 1976; Green et al. 1981).

Burst fractures result from more direct forms of axial compression (Fig. 3.22). Computed tomographic assessment of wedge and burst fractures shows additional fractures of the posterior bony elements which were not identifiable on the plain radiographs even with multiple projections. It has been shown that fractures of the vertebral arch occur with considerably greater frequency than fractures of the vertebral body (Miller et al. 1978). The majority of patients have more than one component to a spinal injury, with four or more components occurring in 16 per cent of patients. In many patients these involve adjacent vertebrae, with significantly more extensive injuries than had been suspected (Hadden and Gillespie 1985) (Fig. 3.23). The extent to which the spinal canal is compromised by bone fragments from a burst fracture cannot be determined reliably without computed tomography. Decisions as to the need for anterior surgical decompression should be

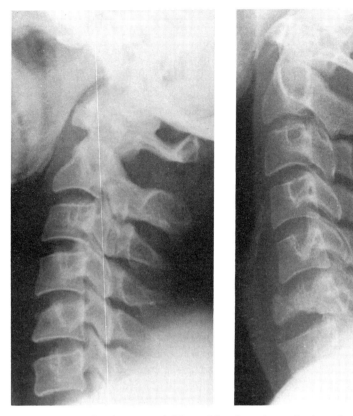

Fig. 3.21. Wedge fracture of C5 resulting from hyperflexion when trampolining.
Fig. 3.22. Burst fracture of C6 sustained in a diving accident.

based on clinical judgement and not primarily upon computed tomographic calculations of the reduction in sagittal diameter or cross-sectional area of the spinal canal, as these correlate poorly with the presence of neurological deficit. On the other hand, the recognition of minor degrees of malalignment and separation due to the soft tissue components of these injuries, which may equally contribute to instability, may be more easily seen on plain radiographs than on computed tomograms. The visualization of such features is less satisfactory in the axial plane and even the best reformatted images lack the discrimination of plain radiographs.

Instability
Computed tomograms confirm that the more extreme unstable categories of burst fracture closely resemble the classic or highly unstable hyperflexion

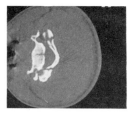
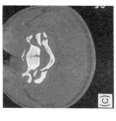

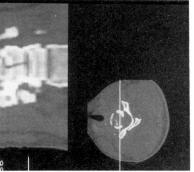
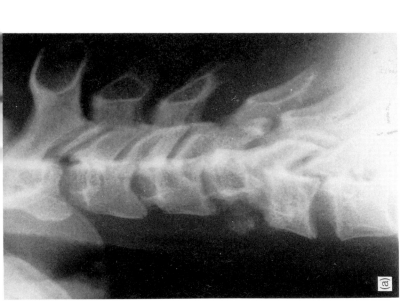

Fig. 3.23. Hyperflexion tear-drop fracture dislocation of C5. The lateral radiograph (a) and sagittal computed tomogram reconstructon (b) show a typical injury with displacement of the larger vertebral fragment posteriorly. The separation of the spinous processes of C4 and C5 (fanning) indicates ligamentous disruption and instability, but the extent of the injuries to the laminae and articular facets shown on an axial computer tomogram (c) was not apparent on the initial radiographs. The spinal canal is not markedly narrowed but the patient, aged 17 years, suffered a total tetraplegia.

tear-drop fracture–dislocation, as suggested in the original description by Schneider and Kahn (1956). In this injury, sustained by compressive hyperflexion, the anterior longitudinal ligament avulses and remains attached to a fragment of the antero–inferior part of the vertebral body. The larger posterior fragment of the body is thus unsupported and free to displace posteriorly with encroachment on the spinal canal (Figs. 3.6 and 3.23). The suggested similarity to burst fractures has been controversial (Harris 1978a,b; Rogers 1981). The three-column concept has clearly established a category of unstable burst fracture with involvement of the middle column. Furthermore, with computed tomography (Scher 1982) burst fractures are identified showing comminution of the upper part of the vertebral body combined with a sagittal split of the remainder. Reformatted sagittal images of such injuries demonstrate a close resemblance to the tear-drop pattern. This injury is the cervical equivalent of the thoracolumbar 'crush–cleavage' fracture described by Lindahl, Willen, Nordall, and Irstam (1983). Its recognition is important as anterior decompression and surgical fusion may be indicated. If the fracture were regarded as a stable burst injury it might be reasonable to wait for interbody fusion to take place spontaneously.

Dislocation of the apophyseal joints

The remaining injury of the hyperflexion type is the bilateral facet dislocation (Figs. 3.1 and 3.24). This results from disruptive hyperflexion. Forces

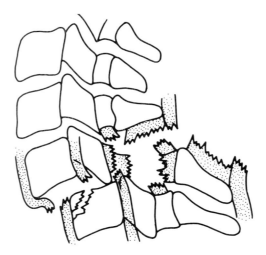

Fig. 3.24. Bilateral facet dislocation. The anterior displacement is greater than one-half of the sagittal diameter of the vertebral body and there is total ligamentous disruption (cf. Fig. 3.1).

Table 3.14. Spinal cord compression in cervical spine injuries

Injury	Frequency of cord compression (per cent)
Bilateral cervical facet dislocation	85
Tear-drop or severe burst fracture	75
Unilateral facet dislocation	30

of distraction and shearing drive the inferior facets of the upper vertebra over the top of the superior facets of the subjacent vertebra where they lock like a ratchet. Such displacement can occur only with total disruption of the intervertebral disc and intervertebral ligaments (Beatson 1963). Fragments of the disc may be driven into the canal. The upper vertebra is displaced anteriorly to the extent of the antero–posterior diameter of the articular pillar. This is always greater than one-half of the diameter of the vertebral body (Beatson 1963). The disruption makes the position totally unstable to further flexion or shearing. There is a higher incidence of damage to the spinal cord than with any other spinal injury (Table 3.14), particularly as the dislocation is more frequently encountered in elderly patients where the spinal canal is narrowed by spondylosis (Braakman and Penning 1971).

Hyperextension injuries

The hyperextension fracture–dislocations classified by Gehweiler as types IV and V (Gehweiler *et al.* 1979) results from more severe hyperextension forces (Figs. 3.7 and 3.25). In extreme hyperextension the spinous processes begin to act as a fulcrum (Forsyth 1964). The head continues to move in an arc imposing anterior distraction and posterior compression forces. These result in rupture of the discovertebral bond and the anterior longitudinal ligament, often with avulsion of a small fragment from the vertebral body. The posterior compression first produces a fracture of the articular pillar (type IV) with subsequent comminution of the pedicles and laminae (type V). The pillar fractures often occur at two adjacent levels and are usually unilateral as a result of some rotation during the injury. Compression of the articular pillar rotates the facets into the horizontal plane so that the apophyseal joint is visible in the frontal view. This also results in loss of superimposition of the articular masses in the lateral view, which is accentuated due to the retropulsion of the posterior fragment. The fracture line is often visible in both planes, extending to the lateral margin and the inferior facet. Such fractures, however, are much better shown on specific 'pillar' views. These require slight rotation of the head in order to avoid

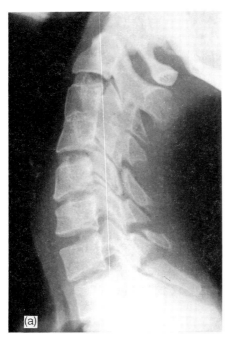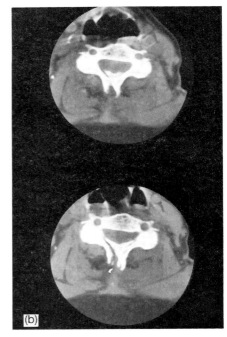

Fig. 3.25. Hyperextension cervical spine injury. There are multiple fractures of the laminae and spinous processes. The C5–6 disc is narrowed, and degenerative lipping subsequently developed at this level within one year. The axial computed tomogram (b) shows the impacted comminuted posterior arches.

superimposition of the mandible, and the presence of an upper cervical spine injury must therefore have been excluded on the basis of earlier radiographs.

In the recoil position the upper vertebra moves forwards with narrowing of the disrupted disc. The position is then similar to that of unilateral facet dislocation. The lack of superimposition of the articular masses in the lateral view may cause further confusion. The frontal projection shows a greater degree of rotation and displacement of the spinous process from the midline following a unilateral facet dislocation. The hyperextension tear-drop fracture results when a larger fragment is avulsed as the anterior longitudinal ligament ruptures. This usually involves the base of the axis, particularly in elderly osteoporotic patients.

Lateral flexion injuries

Forces of lateral flexion are seldom applied in isolation, but usually occur in combination with rotation, axial compression, or hyperextension (Fig.

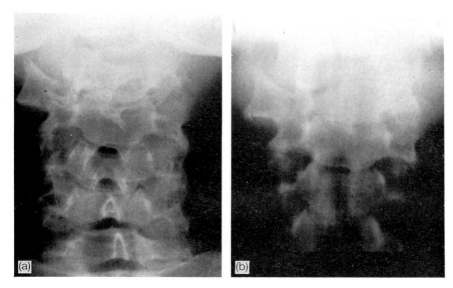

Fig. 3.26. Hyperextension cervical spine injury with lateral flexion. The frontal radiograph (a) and plain computer tomogram (b) show a unilateral injury of the articular pillar with rotation of the facets so that the apophyseal joints are visible.

3.26). Lateral hyperflexion may result in a fracture of the lateral mass of the atlas (Abel and Teague 1979), of the lateral articulation of the axis, or of the articular pillar. It is a component in the production of fractures of the uncinate process or of the transverse process of C7. This results from indirect forces producing avulsion from musculoligamentous pull analogous to those of a clay shoveller fracture of the spinous process. The classical asymmetric injury, in which lateral flexion plays a part, combined with rotation, hyperflexion, and an element of shearing, is unilateral dislocation of a facet joint (Roaf 1963; Schaaf, Gehweiler, Miller, and Powers 1978). This is usually regarded as a stable injury because the overriding facet lodges in the intervertebral foramen. The contralateral articular pillar confers further stability and limits the anterior displacement of the vertebral body to less than half the antero–posterior diameter. Beatson (1963) has identified the severe disc and ligamentous damage that has to take place before unilateral dislocation can occur. In the presence of such disruption it is unlikely that the injury can be regarded as fully stable in respect of further rotation in the direction of the dislocation. The radiographic position resembles that of recoil from a hyperextension injury (Figs. 3.2 and 3.27). There is a greater degree of rotation and tilting so that the disrupted disc widens in the frontal projection and the intervertebral foramen appears enlarged in the oblique view. Neurological injuries occur in about 30 per

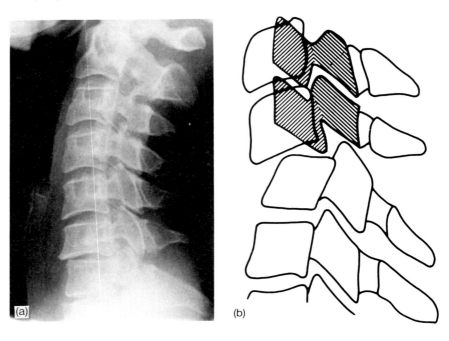

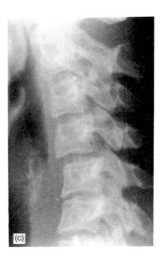

Fig. 3.27. Unilateral facet dislocation. The lateral radiograph (a) and diagram (b) show the loss of superimposition of the articular pillars above the level of the injury. This obliquity distinguishes the position from a hyperextension injury (c) in which the recoil position can otherwise be similar (cf. Fig. 3.2).

cent of patients, usually in the form of a hemicord (Brown Séquard) syndrome or a root lesion.

Pre-existing abnormalities of the cervical spine

Conditions altering the biomechanics of the cervical spine may determine the level or pattern of a subsequent injury. This results most commonly when the range of movement of the lower cervical spine is limited by spondylosis. Similar factors may operate in diffuse idiopathic skeletal hyperostosis (DISH) or following interbody fusion, whether the latter be congenital or acquired, as a result of previous infection or trauma. The most striking examples are seen in ankylosing spondylitis (Fig. 3.28) where the fused, osteoporotic spine may fracture like a stick of chalk after a minor injury. The ossification in the annulus makes the fracture equally likely to pass through the disc or the vertebral body, the latter being more likely to result in neurological damage. This appears to be independent of the mechanism of injury or the extent of displacement and may be related to the

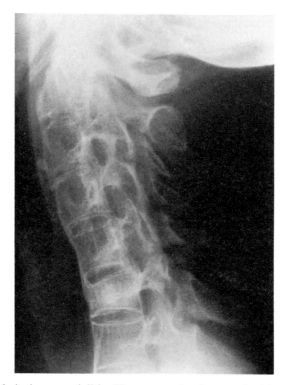

Fig. 3.28. Ankylosing spondylitis. Hyperextension has resulted in a fractured C6 facet. Note the previous wedge compression of the body of C6.

greater incidence of spinal epidural haematoma (Guttman 1966; Harding, McCall, Park, and Jones 1985).

Delayed myelopathy

In contradistinction to atlanto–axial injuries the development of a delayed myelopathy following an injury to the lower cervical spine is seldom the result of instability. It usually follows the processes of repair, when spontaneous interbody fusion in a slightly malaligned position causes further encroachment on the spinal canal. The cord is compressed within the narrowed segment and the mechanisms and clinical features resemble those of cervical spondylosis. There is indeed a predilection for degenerative

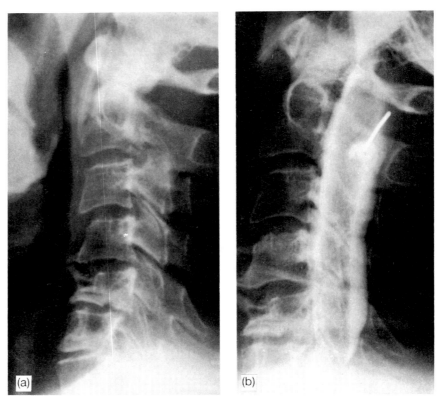

Fig. 3.29. Delayed cervical myelopathy. This 58-year-old man sustained fractures of C5 and C6 playing rugby in 1947. He made a complete recovery from transient tetraparesis. In 1980 he developed signs of progressive cervical cord compression. The plain lateral radiograph (a) shows spondylotic narrowing of the spinal canal superimposed on the old injury. The cervical myelogram (b) shows a complete extradural block to contrast medium.

spondylosis to involve previously injured motion segments, sometimes within a surprisingly short time and following an injury that originally showed no radiological abnormality. The neurological deterioration is not seen until these changes become superimposed upon those of the injury (Fig. 3.29) Similar processes of post-traumatic arthropathy may affect an injured facet joint and be responsible for radicular symptoms also resembling those of degenerative disease.

Myelography

Myelography and computed tomography will establish the need for surgical decompression, usually laminectomy, for which there is no contraindication, as post-traumatic fusion has already stabilized the neck. The value of myelography and computed tomographic myelography, in the investigation of acute brachial plexus injuries, has been overstated. The procedure is often performed with great difficulty in a patient with multiple injuries, and the paralysed arm is extremely painful. The visualization of avulsed nerve roots and post-traumatic avulsion cysts (Fig. 3.30) (Davies, Sutton, and Bligh 1966) is of some value prior to microsurgical techniques, but much of the information can be derived from nerve conduction studies. The myelographic appearances of cystic myelopathy are discussed elsewhere.

Autonomic dysfunction

Injuries to the spinal cord above the thoracolumbar sympathetic outflow may produce autonomic imbalance. The unopposed vagal activity may lessen cardiorespiratory reserve. The use of high ionic contrast media for intravenous urography, arteriography, or venography may cause arrhythmias or cardiac arrest. This risk is avoided by the use of low ionic media, and there is the additional benefit of reducing the incidence of vomiting or thrombophlebitis. Tilting the X-ray table, for example, during myelography may embarrass ventilation that is being maintained solely by the diaphragm, particularly when the abdomen is distended due to paralytic ileus. The latter also increases the risk of regurgitation and aspiration. Most commonly, reflex sympathetic hyperactivity is caused by distension of the bladder during cystourethrography. This is responsible for hypertension and a pounding headache, together with sweating and flushing at levels above the cord lesion.

Thoracolumbar injuries

The major injuries and fracture–dislocations of the thoracic and lumbar spine are listed in Table 3.15 and illustrated in Fig. 3.31(a)–(f).

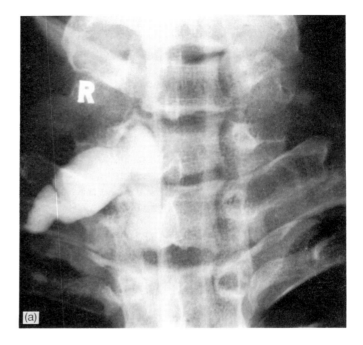

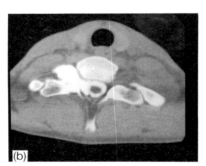 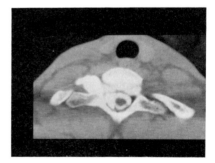

Fig. 3.30. Brachial plexus avulsion. Myelography (a) and computed tomography (b) show avulsed nerve roots and post-traumatic avulsion cysts six weeks after a motor cycle accident.

Introduction—vertebral compression

The range of movement of the thoracic spine is restricted by the rib cage, and the lumbar spine is buttressed by the soft tissues of the paraspinal muscles and the abdomen. The normal dorsal kyphos predisposes to anterior wedge compression of the vertebral bodies, because forces of axial compression are converted into those of flexion. Most fractures of dorsal vertebrae, whether

Table 3.15. Injuries of the thoracolumbar spine

Wedge fracture	—stable
	—unstable
Burst fracture	—stable
	—unstable
Distraction fracture (seat belt)	
Rotary slice fracture–dislocation	
Shear fracture–dislocation	
Hyperextension fracture–dislocation	
Isolated direct posterior fractures	
Avulsion fractures	

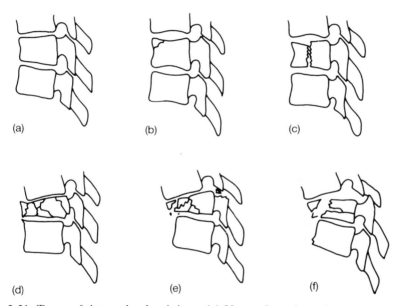

(a)　　　(b)　　　(c)

(d)　　　(e)　　　(f)

Fig. 3.31. Types of thoracolumbar injury. (a) Upper thoracic wedge compression. (b) Lower thoracic burst compression. (c) and (d) Increasing comminution and displacement into the spinal canal. (e) Rotary slice fracture–dislocation. (f) Shear fracture–dislocation.

occurring spontaneously or as the result of more identifiable trauma, are therefore of the wedge-compression type. The incidence of such fractures increases with age, particularly in relation to osteoporosis. The reduced mineral content of the vertebral body, and the declining elasticity of the body and the intervertebral disc, often result in multiple fractures of the

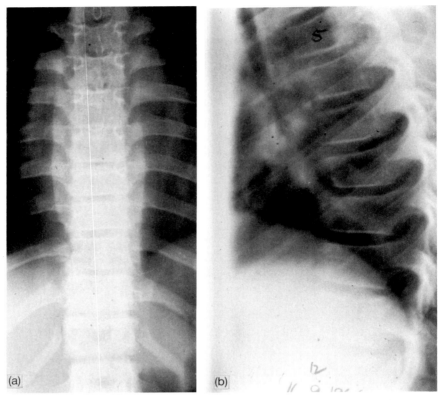

Fig. 3.32. Dorsal wedge compression fractures. The frontal radiograph (a) shows a paravertebral soft tissue shadow. The lateral view (b) shows multiple compression fractures sustained in a fall from a tree.

mid- and lower dorsal spine after minimal trauma. Nicholas, Wilson, and Freiberger (1960), however, showed that fractures occurring spontaneously in the rarefied spine proved to be due to metastatic disease in 20 per cent of patients over the age of fifty-five years and in 50 per cent of patients under this age. Compared with osteoporotic collapse, metastatic disease is much less likely to show compression of multiple vertebral bodies. There are few indirect radiographic signs of thoracolumbar vertebral injury, but a paraspinal haematoma [Fig. 3.32(a)] may produce a fusiform soft tissue density. In the thorax this may extend over the lung apex with a similar radiographic appearance to that seen with a torn aorta. Evidence of retroperitoneal haematoma causing displacement or blurring of the outlines of the soft tissues in the presence of a lumbar vertebral injury is much less common.

Biomechanics

Compression fractures resulting from more identifiable trauma occur more commonly at the thoracolumbar junction, particularly involving the twelfth thoracic and first lumbar vertebral bodies. This is a zone of mechanical transition between the relatively immobile thoracic spine and the more flexible lumbar spine. In the erect position the spinal curves reverse at this level, but when the back is flexed, as following a protective reflex, it is the site of greatest anterior curvature. The facets show an abrupt change in orientation between the thoracic and the lumbar spine. The so-called mortice joint which resists all movements other than flexion (Davis 1955) is usually between T11 and T12. Resistance to rotational forces is therefore concentrated on the thoracolumbar junction. The smaller increase in injuries at the cervicodorsal junction is the result of similar mechanical factors. There are further biomechanical analogies between cervical and thoracolumbar injuries, in that stability depends on the integrity of the spinal ligaments, and it has been shown that forces other than those of simple flexion or extension—rotation, shearing, or distraction—are responsible for ligamentous damage (Roaf 1960; Holdsworth 1963).

Radiology

Thus for thoracolumbar injuries, as for cervical ones, radiology has to answer questions of vital clinical importance: Is the injury stable? Is the spinal canal compromised? Decisions regarding the need for surgical reduction and stabilization, or operative decompression of the spinal canal, will be based on a radiological as well as a clinical evaluation. The radiographic projections and techniques available for examination of the lumbar spine have been listed in Table 3.2. The radiological signs of instability are shown in Table 3.16. The bulk of the trunk makes the radiological demonstration and evaluation of these features more difficult in the lumbar spine. Greater reliance is placed on the antero–posterior projection than for cervical spine injuries, and inspection of the neural arches and spinous processes on the lateral radiograph requires a bright light. In the past, plain radiography has frequently been supplemented by tomography. Computed tomography scanning has considerable advantages over conventional tomographic techniques; these are listed in Table 3.17. The disadvantages (Table 3.18) are

Table 3.16. Radiographic signs of instability in lumbar spine injuries

Disruption of intervertebral disc
Disruption of an apophyseal joint or a facet fracture
Separation of laminae or spinous processes on lateral radiograph
Malalignment of pedicles or spinous processes on frontal radiograph

Table 3.17. Advantages of computed tomographic scanning in spinal injuries

---
Less manipulation of the patient
Lower radiation dosage
Better visualization of —spinal canal in the axial plane
      —all the components of the injury
      —difficult spinal levels
      —intraspinal and paraspinal soft tissues
Assessment of trauma to other systems

Table 3.18. Disadvantages of computed tomographic scanning in spinal injuries

---
Cost and accessibility
Decreased detail compared with conventional tomography
Poor visualization of —fractures in the horizontal plane
      —vertebral compression or translocation
      —posterior ligamentous disruption
Artefact from external fixation

---

less significant and their importance is dwindling as scanning techniques improve. Nevertheless, the role of plain radiography remains paramount: computed tomography should never be performed as the sole investigation or as a screening technique, and it should be reserved for detailed study of a limited segment and should not be expected to discriminate intraspinal soft tissue pathology. Displacement in the horizontal plane is easily overlooked on successive axial sections (Handel and Lee 1981), and some of the aspects of posterior ligamentous disruption and instability are better shown on conventional radiographs.

Wedge and burst fractures

In the common flexion–compression injury the fulcrum passes through the nucleus *pulposus* and the resulting forces of posterior distraction are relatively small. The vertebral body, consisting of compressible cancellous bone within a thin cortex, is subjected to maximal compression forces at its anterior margin. The intervertebral discs at upper dorsal levels are thin and relatively non-compressible. The anterior cortex of the vertebral body is therefore likely to fracture before the end-plate is compressed, giving rise to a true wedge configuration [Fig. 3.22(b)]. Even with the extended modern concepts of stability, such fractures are seldom regarded as unstable (Kilcoyne, Mack, King, Ratcliffe, and Loop 1983). Stability should not be assumed, however, when there is a fracture, separation, or malalignment of the posterior bony arches.

The lower dorsal and lumbar discs are thicker. Mounting pressures within the nucleus bow and fracture the vertebral end-plate after expressing blood from the cancellous bone of the vertebral body (Roaf 1960). This gives rise to a biconcave or burst configuration. The anterior cortex collapses only as higher pressures are reached. The dense cortical bone of the neural arch and articular processes is more resistant to compression but will ultimately fracture. If the compression force is not accompanied by flexion, the explosion of disc material into the vertebral body may produce a complete split in the coronal plane. More commonly there is comminution, predominantly of the upper half of the vertebral body, with a sagittal split more inferiorly—the crush–cleavage fracture (Lindahl *et al.* 1983). This pattern may be appreciated only by computed tomography (Kilcoyne *et al.* 1983; Willen, Lindahl, Irstam, Aldman, and Nordwall 1984). Fragments of the vertebral body are displaced anteriorly or driven posteriorly into the spinal canal (Fig. 3.33).

Instability of wedge and burst fractures

These fractures, whether of wedge or burst type, have been regarded in the past as stable, as there seemed to be little posterior distraction or destruction of the ligamentous complex (Nicoll 1949; Holdsworth 1963). Whitesides (1977) recognized that some of these injuries showed extensive ligamentous disruption, and computed tomographic scanning (Kilcoyne *et al.* 1983; Willen *et al.* 1984) has confirmed that fractures of the pedicles and laminae are almost invariably present, together with separation and subluxation of the apophyseal joints. McAfee *et al.* (1982) and Denis (1983) have considered as unstable those fractures which result in a deformity that may progress under physiological loads or in which there is evidence of damage to the spinal cord or *conus* with a potential for deterioration in the neurological status. Within the three-column theory such fractures, particularly the burst fractures of the crush–cleavage type, have sustained damage to the middle as well as to the anterior column. The role played by computed tomography in the elucidation of an apparently simple burst fracture is illustrated in Fig. 3.33.

Distraction fractures

In fractures of the seat-belt type (Figs. 3.34 and 3.35) the fulcrum is transferred to the anterior abdominal wall so that much higher forces of posterior distraction are generated (Roaf 1960). The patterns of injury range from the purely discoligamentous to the Howland (Howland, Curry, and Buffington 1965) fracture in which the vertebral body and neural arch suffer a total horizontal split and separation (Fig. 3.34). The common

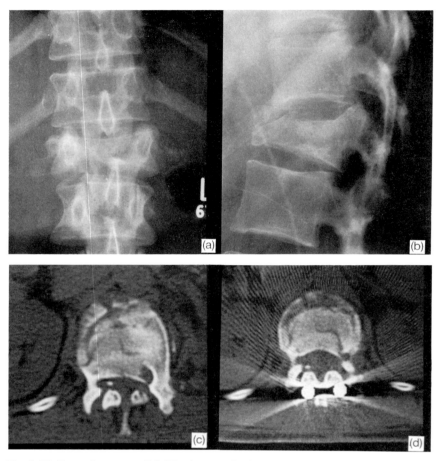

Fig. 3.33. Burst fracture at L1. The interpedicular distance is increased, with loss of alignment of the pedicles and spinous processes (a). Disruption of the facet joints and bursting of fragments posteriorly into the spinal canal is better shown on axial computed tomography [(c) and (d)] made before and after stabilization.

patterns of injury (Chance 1948; Smith and Kaufer 1969) show horizontal fractures of the vertebral body, parallel to the superior end-plate, emerging through the interspinous ligament or the spinous process. The articular facets fracture and the apophyseal joints sublux. The extent of damage to the posterior column renders these fractures unstable. About 15 per cent of patients show evidence of cord damage (Rogers 1971), which is likely to be aggravated if the nature of the injury is not recognized, and the application of distraction rods results in further separation. The radiological demonstra-

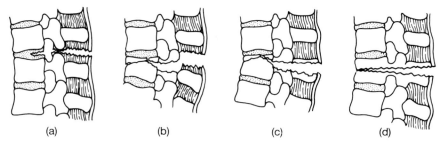

Fig. 3.34. Types of lumbar distraction fracture. (a) Ligamentous injury. (b) Smith fracture. (c) Chance fracture. (d) Howland fracture.

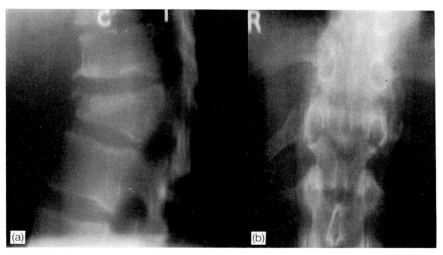

Fig. 3.35. Seat-belt fracture at D12. This injury was sustained by a front-seat passenger in a road traffic accident. The lateral radiograph suggested a wedge fracture, but the frontal view (a) and computer tomogram (b) show the fracture line passing horizontally through the pedicles into the laminae.

tion of a thoracolumbar injury of the distraction type should always lead to a consideration of the possibility of other seat-belt injuries—to the abdominal wall, the gut, or a solid viscus—the seat-belt syndrome (Table 3.19) (Garret and Braunstein 1962; Hampson, Coombs, and Hemingway 1984; Newman 1984)—which are present in about 30 per cent of patients.

Rotary slice fracture–dislocation
In addition to predisposing towards a bursting element in compression fractures, the distortion of the resilient lower dorsal and lumbar discs converts the forces of compression and flexion into those of rotation and

Table 3.19. The seat-belt syndrome (Garret and Braunstein 1962)

Abrasions and hernias of abdominal wall
Longitudinal splits of antimesenteric border of small bowel
Avulsion of outer layers of sigmoid colon
Rupture of duodenum or pancreas
Rupture of liver, spleen, kidney, gravid uterus
Transverse fractures of the lumbar spine with injury to the cord or *cauda equina*

shearing, like a soft tyre when a car is turning (Fig. 3.36). At low levels of energy this may result only in some damage to the disc itself or the detachment of a small chip from the vertebral margin. The more rigid superior articular processes force the inferior processes medially and posteriorly. The articular surfaces are then damaged, with the subsequent development of a traumatic arthritis and pain radiating to the *gluteus medius*, greater trochanter, and iliotibial band, and thence to the upper lateral calf—the facet syndrome (Farfan 1984). The torsional deformity of the outer layers of the annulus is greatest at the posterolateral angles of the disc, where the emerging nerve roots may be compressed. The iliolumbar ligaments are strong enough to protect the lumbosacral level and the effects are most frequently seen at L4–5 (MacGibbon and Farfan 1979). More severe rotational forces make their impact at the thoracolumbar junction. The vertebral rim is then avulsed by the densely attached fibres of the annulus. Increasing rotational and shearing forces split larger fragments from the vertebral body parallel to the superior end-plate. These may

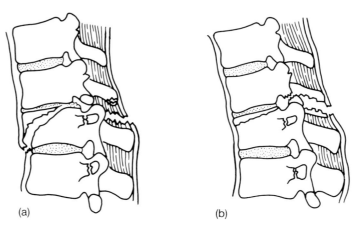

(a)   (b)

Fig. 3.36. Rotary slice fracture–dislocation of the thoracolumbar spine. (a) With posterior ligamentous disruption and apophyseal dislocation. (b) With fractures of an inferior articular process and the posterior bony arch of the upper vertebra.

separate and displace into the spinal canal, where fragments of less than 1 mm in diameter have been shown experimentally by computed tomography (Lindahl, Willen, and Irstam 1983). The longitudinal and posterior ligaments tear and the apophyseal joints fracture, sublux, or dislocate. The lumbar apophyseal joints have stronger capsules and a smaller range of movement than the cervical. Much greater forces are therefore required to produce disengagement, and dislocation is unlikely to occur without an associated fracture. With the production of a rotary slice fracture–dislocation of this type [Fig. 3.36(a), (b)], the incidence of cord damage rises dramatically, ranging from 53 to 70 per cent in different series. Bohler (1956) noted that neurological injuries were less severe when fractures of the posterior bony arch were present, describing this component of the injury as the 'salvation of the cord'. Holdsworth (1963) and many subsequent authors have noted the extent to which the injury may be reduced in the supine position at the time of radiological examinations. The presence of oblique or horizontal fractures of the transverse processes, especially if these be unilateral, or minor degrees of apophyseal separation or fractures of the articular facets, may be indications of a more severe rotational injury than that which at first seems apparent.

Shear fracture–dislocation

Injuries with a more marked horizontal component, usually transmitted from a direct blow posteriorly, may not only fracture the vertebral body but may also produce complete shearing of the discovertebral bond and translocation of the upper spine upon the lower [Fig. 3.37(a), (b)]. At the moment of injury the spinal canal is compromised and malaligned, with the production of a severe neurological injury. Ligamentous disruption is usually total however, so that, as with the rotary injuries, the alignment of the spine may return almost to normal in the supine position. The bony neural arch may

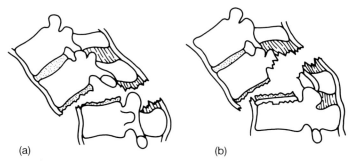

(a) (b)

Fig. 3.37. Shear fracture–dislocation of the thoracolumbar spine. (a) With apophyseal dislocation. (b) With fractures of the inferior articular process of the upper vertebra.

remain intact, with posterior injuries limited to fractures of the articular processes or apophyseal dislocation (Fig. 3.37). More extensive posterior fractures produce a traumatic spondylolisthesis, with some reduction in the frequency and severity of the damage to the cord or *conus* (Bohler 1956).

Hyperextension injuries

Hyperextension fracture–dislocations of the thoracolumbar spine are uncommon and are illustrated in Fig. 3.38 (a)–(d). Burke (1971*b*) described only four such patients in a series of 154 thoracolumbar injuries. All of these involved the thoracic spine and were associated with severe neurological injuries. Roaf (1960), working with cadaver spines, showed that pure hyperextension forces produced only posterior compression fractures of the vertebral arches and spinous processes and that disruption of the dis-covertebral bond and anterior longitudinal ligament would not occur without a rotation or shearing force. The lateral radiograph shows anterior distraction or fission fractures of the vertebral body, sometimes with a small compression fracture of the postero–inferior corner. The position is highly unstable and the prognosis is poor. The fractures seen in ankylosing spondylitis are not infrequently of this type (Guttman 1966). Isolated fractures of the posterior bony arch or the articular processes, sustained as a result of direct injuries, are rare, and most fractures of this type are combined with fracture–dislocation of the rotary or shear type. Fractures of the transverse or spinous processes, however, are more common and, when resulting from muscular avulsion, are usually orientated vertically.

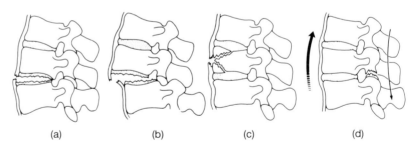

| (a) | (b) | (c) | (d) |

Fig. 3.38. Hyperextension fracture–dislocation of the thoracolumbar spine: with disc and anterior ligamentous disruption and apophyseal separation (a) or with dislocation (b); with a fission fracture of the anterior part of the vertebral body (c). A posterior compressive force may fracture the bony arch (d).

Radiology of fracture–dislocations

Examples of rotary slice and shear fracture–dislocations are shown in Figs. 3.39, 3.40, and 3.41. It is evident from these that neither stability, nor encroachment of bony fragments into the spinal canal, can always be reliably

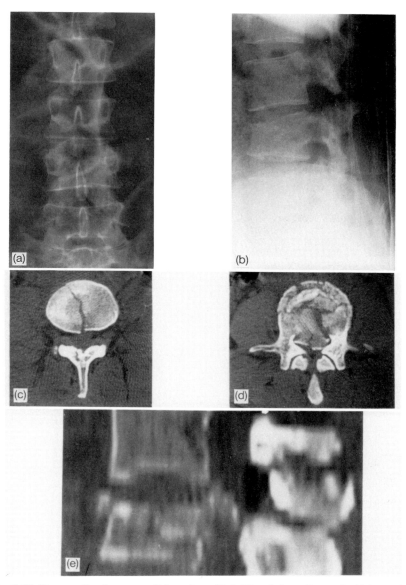

Fig. 3.39. Rotary slice fracture–dislocation of L4. The plain radiographs [(a) and (b)] made after a riding accident appear to show a wedge fracture. The extension of the fracture through the spinous process of L3 (cf. Fig. 3.36) indicates posterior disruption resulting from a rotational component in the injury. The posterior cortex of the compressed vertebral body is clearly shown on the lateral radiograph (b), with no indication of the extent to which the canal is narrowed. Axial computed tomography [(c) and (d)] shows that the fracture is of the crush–cleavage type and that the spinal canal is almost obliterated. A reformatted sagittal image (e) shows that the bone fragments are at the level of the pedicles and are therefore obscured on the lateral radiograph.

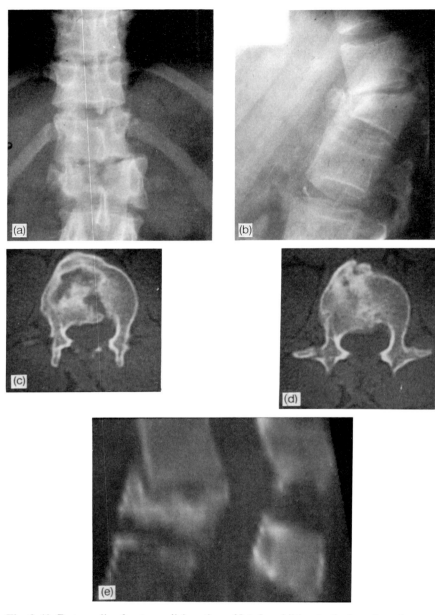

Fig. 3.40. Rotary slice fracture–dislocation of L1. In addition to the bursting element the posterior disruption has resulted in an angular kyphos with separation of the spinal processes [(a) and (b)]. Computed tomography [(c) and (d)] shows that the superior facets of L1 are 'empty', indicating apophyseal separation. This is an unstable injury. The spinal canal is not seriously narrowed (e).

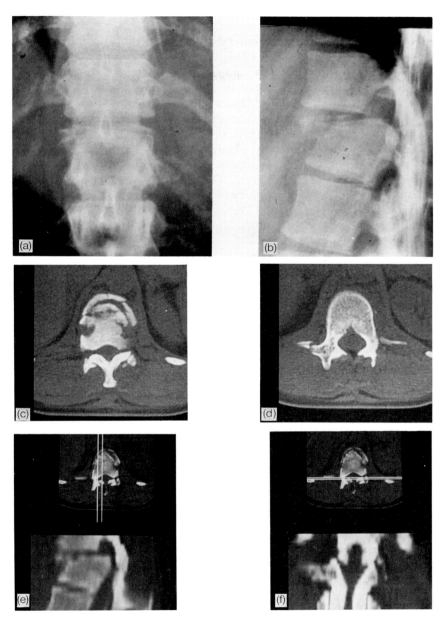

Fig. 3.41. Shear fracture–dislocation of L1. The separation of the spinous processes (a) has given rise to the appearance of an 'empty vertebra'. The shearing component is greater than in Fig. 3.40 and has resulted in total bilateral dislocation of the facet joints. Axial and reformatted computed tomography shows the inferior facets of D12 lying within the spinal canal, which is therefore severely narrowed [(c)–(f)].

assessed from plain films. Computed tomography provides an indispensable guide to surgical management, either by operative reduction and stabilization or by anterior surgical decompression and removal of bone fragments encroaching on the spinal canal (Brandt-Zawadski, Miller, and Federle 1981; Handelberg, Bellemans, Opdecam, and Casteleyn 1981; Donovan Post, Green, Quencer, Stokes, Callahan, and Eismont 1982). Indications for surgery, nevertheless, remain primarily clinical and are based on deteriorating neurological status. As with cervical injuries, decisions should not depend on computed tomographic measurements of the extent to which the spinal canal is compromised. The poor correlation between such measurements and the severity of a neurological deficit has been noted by McAfee *et al.* (1982) and Denis (1983). The role of computed tomography in the management of spine and cord injuries is summarized in Table 3.20.

Table 3.20. Computed tomography in spinal trauma

Indications

Any injury resulting in
   A significant spinal fracture or dislocation
   A neurological deficit

Purpose

More accurately to assess spinal stability and the extent to which the spinal canal is compromised by bone fragments

This provides a guide to surgery

Reduction and stabilization
Anterior decompression

Post-operative examination

Plain radiographs or computed tomographic scans, performed after early reduction and stabilization of a severe burst injury or thoracolumbar fracture–dislocation, will demonstrate the extent to which the anterior vertebral height is regained and the normal posterior cortical alignment is restored (Figs. 3.42 and 3.43) (White, Newberg, and Seligson 1980). Experimental studies indicate that a component of the overall neurological loss may be caused by secondary injury and suggest that early post-traumatic decompression may reduce post-traumatic oedema and improve the blood supply to the injured cord. It has been difficult, however, to define the clinical aspects that indicate that a patient will benefit from surgery (Collins and Chehrazi 1982). The main purpose, therefore, of reduction and stabilization is to lessen the risk of further damage to the cord, to correct or reduce the spinal deformity, to make nursing easier, and to shorten the stay

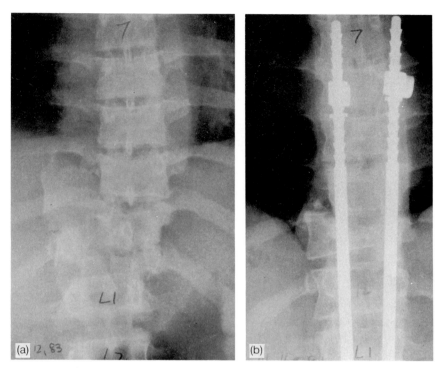

Fig. 3.42. Shear fracture–dislocation of D11–12. Frontal radiographs show spinal realignment following reduction and stabilization (Mr R. Jackson). The patient suffered no neurological injury.

in hospital. This helps to prevent respiratory, urinary, and cutaneous complications (Kaufer and Hayes 1966; Lewis and McKibbin 1974) and facilitates early rehabilitation, which makes the most of the retained neurological functions and is so important psychologically. In the severely injured patient it may not be possible to reduce and stabilize the spine at an early stage. It is then unlikely that radiological re-alignment of the spine and spinal canal will be achieved. In the presence of an incomplete neurological injury, bone fragments encroaching on the spinal canal are then removed via an anterior approach and the adequacy of such removal can be confirmed by computed tomography. The development of a delayed and possibly progressive kyphosis, following thoracolumbar injury, represents failure of the initial management. Apart from the deformity, this may cause pain, a deterioration in the neurological status, and difficulties with rehabilitation (Roberson and Whitesides 1985). It is particularly likely to be seen if a laminectomy has been performed as the initial decompressive procedure.

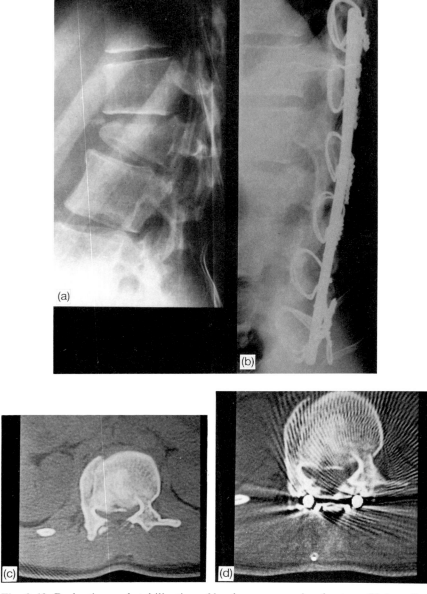

Fig. 3.43. Reduction and stabilization of lumbar compression fracture. Plain radiographs [(a) and (b)] and axial computed tomograms [(c) and (d)] made before and after insertion of Harrington distraction rods (Mr R. Jackson) show decompression of the spinal canal and restoration of vertebral height and alignment.

Laminectomy either has no effect on the spinal cord injury or increases the morbidity (Bedbrooke 1975). It has been suggested that the posterior displacement of the spinal cord, shown myelographically at the level of a laminectomy, reduces the anterior and/or radicular blood supply (Bennett and McCallum 1957).

Myelography

The value of myelography in the investigation of patients with acute cord trauma is closely related to perceptions about the role of surgery. Many centres in the USA regard myelography as valuable and mandatory, in providing an indication for immediate surgical decompression, in all patients showing a neurological deficit. Guttman (1973) stated that 'in the vast majority of vertebral injuries the damage to the spinal cord is instantaneous', and it is widely held in Europe that neurological loss following injury results from intrinsic cord damage that cannot be altered by surgery. Although it has been shown in animal experiments (Freeman and Wright 1985) that early myelotomy may protect the traumatized spinal cord, there is no evidence that this technique alters the effects of human spinal cord injury (Collins and Chehrazi 1982). At the Wessex Neurological Centre and the Wessex Spinal Injuries Unit myelography is therefore regarded as playing a very restricted role in the examination of patients with injuries to the spine and spinal cord. A myelographic block is almost invariably present prior to the reduction of a fracture–dislocation; its demonstration adds nothing to the surgical management, as most blocks are relieved by reduction and stabilization. Myelography constitutes a further insult to the damaged cord, both through the manipulation involved in the performance of the procedure and through the introduction of contrast medium, which is sometimes seen to pass into the damaged cord (Dubois, Drayer, Sage, Osborne, and Heinz 1981). Conner, Martuza, and Heros (1982) reported seizures following post-traumatic myelography using metrizamide. The specific indications for acute myelography are listed in Table 3.21. The additional role in patients with injuries to the brachial plexus, or delayed myelopathy of the spondylotic type, following cervical injuries has already been mentioned. Post-traumatic avulsion cysts of lumbar nerve roots have been reported, but are rare (Streiter and Chambers 1984). Myelography is also of value in the investigation of post-traumatic myelopathy due to the

Table 3.21. Indications for immediate myelography in spinal injuries

Evidence of injury to the spinal cord or nerve roots, with no corresponding
    fracture or dislocation
Neurological deterioration after reduction and stabilization of the spinal injury
Development of a neurological deficit at a level remote from the primary injury

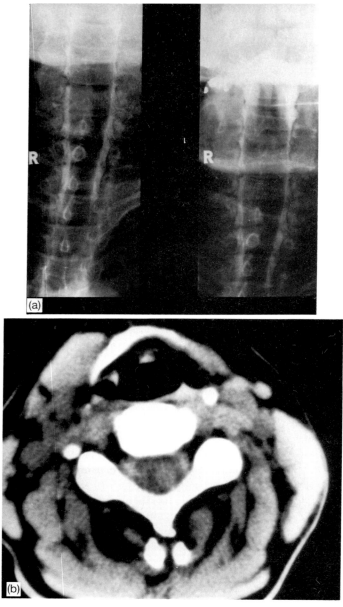

Fig. 3.44. Post-traumatic cystic myelopathy. Cervical myelography (a) shows an expanded cervical and upper dorsal cord which changed size between the prone and supine positions. Computed tomography (b) performed four hours after the myelogram shows contrast medium within the cystic area.

extension of fluid-filled cavities or cysts along the length of the spinal cord by the process of cavitation necrosis. The result may be the formation of several large fluid-filled cavities or many small cavities. The volume of the cavities adjacent to the injury may be substantial, leaving only small bridges of tissue. Progressive cavitation may cause the loss of remaining or recovered neurological status or autonomic dysfunction (Guttman 1973). The classical myelographic sign (Fig. 3.44) is that of an expanded cord, but this is present in only about half the patients showing typical clinical features, with a substantial minority showing a normal-sized or small atrophic cord (Stevens, Olney, and Kendall 1985). The distinction between a large cyst and multiple small cysts is more difficult, but the former can be inferred when changes from the prone to the supine position alter the size of the cord. Such alterations may be better shown on post-myelographic computed tomography when, in addition, if the examination be deferred for several hours, contrast medium may be shown to accumulate within the substance of the

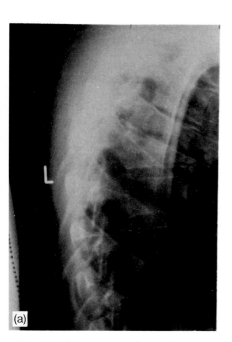
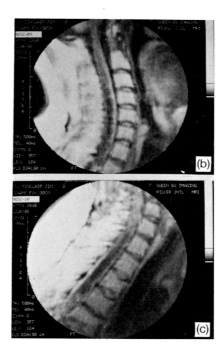

Fig. 3.45. Post-traumatic cystic myelopathy. The plain radiograph (a) shows a mild dorsal fracture–dislocation sustained two years previously in a road traffic accident. Magnetic resonance images [(b) and (c), courtesy of the Queen Square Imaging Centre], obtained when the patient showed an ascending neurological deficit, show low signal intensity from an intramedullary post-traumatic cyst extending from the level of the injury to the axis vertebra.

cord (Seibert, Dreisbach, Swanson, Edgar, Williams, and Hahn 1981; Quencer, Green, and Eismont 1983). Distinction between the cystic and non-cystic types of myelopathy is important if a cyst-subarachnoid or cyst-peritoneal shunt procedure is to provide effective decompression and prevent the inexorable extension of the cyst along the length of the cord, possibly even achieving some return of newly lost function (Mudge, van Dolson, and Lake 1984; Stevens *et al.* 1985). Ultrasonography of the exposed cord at operation will provide information as to the best siting of the shunt catheter, but pre-operative evaluation is even better achieved by magnetic resonance imaging (Fig. 3.45).

Acknowledgements

I acknowledge with thanks the assistance of Miss Claire Styles, Wessex Neurological Centre, who typed the manuscript with many revisions, and Mr David Whitcher and Mr Peter Jack, Department of Teaching Media, Southampton General Hospital, for their photography.

References

Abel, M. S. and Teague, J. H. (1979). Unilateral lateral mass compression fractures of the axis. *Skeletal Radiol.* **4**, 92–8.

Alker, G. J., Oh, Y. S., Leslie, E. V., Lehotay, J., Panaro, V. A., and Eschner, E. G. (1975). Postmortem radiology of head and neck injuries in fatal road traffic accidents. *Radiology* **114**, 611–17.

Anderson, L. D. and d'Alonzo, R. T. (1974). Fractures of the odontoid process of the axis. *J. Bone Joint Surg.* **54**(A), 1663–74.

Apuzzo, M. L. J., Heiden, J. S., Weiss, M. H. (1978). Acute fractures of the odontoid process. An analysis of 45 cases. *J. Neurosurg.* **48**, 85–91.

Barnes, R. (1948). Paraplegia in cervical spine injuries. *J. Bone Joint Surg.* **30**(B), 234–44.

Beatson, T. R. (1963). Fractures and dislocations of the cervical spine. *J. Bone Joint Surg.* **45**(B), 21–35.

Bedbrooke, G. M. (1975). Treatment of thoracolumbar dislocation and fractures with paraplegia. *Clin. Orthoped. Related Res.* **112**, 27–43.

Bennett, M. H. and McCallum, J. E. (1977). Experimental decompression of spinal cord. *Surg. Neurol.* **8**, 63–7.

Bohler, L. (1956). *The treatment of fractures.* Grune and Stratton, New York.

Braakman, R. and Penning, L. (1971). *Injuries of the cervical spine.* Excerpta Medica Foundation, Amsterdam.

Brandt-Zawadski, M., Miller, E. M., and Federle, M. P. (1981). C.T. in the evaluation of spine trauma. *Am. J. Roentgenol.* **136**, 369–75.

Brunton, F. J., Wilkinson, J. A., Wise, K. S. H., and Simonis, R. B. (1982). Cine radiography in cervical spondylosis as a means of determining the level for anterior fusion. *J. Bone Joint Surg.* **64**(B), 399–404.

Burke, D. C. (1971*a*). Spinal cord trauma in children. *Paraplegia* **9**, 1–14.

Burke, D. C. (1971*b*). Hyperextension injuries of the spine. *J. Bone Joint Surg.* **53**(B), 3–12.

Calenoff, L., Chessare, J. W., Rogers, L. F., Toerge, J., and Rosen, J. S. (1978). Multiple level spinal injuries: importance of early recognition. *Am. J. Roentgenol.* **130**, 655–9.

Cattell, H. S. and Filtzer, D. L. (1965). Pseudosubluxation and other normal variations in the cervical spine in children. *J. Bone Joint Surg.* **47**(A), 1295–309.

Chakera, T. M. H., Anderson, J. E. M., and Edis, R. H. (1982). Atlanto–axial dislocation and vertebral artery aneurysm. *Br. J. Radiol.* **55**, 863–4.

Chance, G. Q. (1948). Note on a type of flexion fracture of the spine. *Br. J. Radiol.* **21**, 452–3.

Cheshire, D. J. (1969). The stability of the cervical spine following the conservative treatment of fractures and fracture dislocations. *Paraplegia* **7**, 193–203.

Clark, W. M., Gehweiler, J. A. Jr., and Laib, R. (1979). Twelve significant signs of cervical spine trauma. *Skeletal Radiol.* **3**, 201–5.

Collins, W. F. and Chehrazi, B. (1982). Concepts of the acute management of spinal cord injury. In *Recent advances in clinical neurology IV* (ed. W. B. Mathews and G. H. Glaser), pp. 67–82. Churchill Livingstone, Edinburgh.

Conner, E. S., Martuza, R. L., and Heros, R. C. (1982). Seizures due to metrizamide myelography in patients with acute cervical spine injury. *Neurosurgery* **10**, 266–7.

Davies, E. R., Sutton, D., and Bligh, A. S. (1966). Myelography in brachial plexus injury. *Br. J. Radiol.* **39**, 362–71.

Davis, P. R. (1955). The thoracolumbar mortice joint. *J. Anat.* **39**, 370–7.

Denis, F. (1983). The three column soine and its significance in the classification of acute thoracolumbar spine injuries. *Spine* **8**, 817–31.

Donovan Post, M. J., Green, B. A., Quencer, R. M., Stokes, N. A., Callahan, R. A., and Eismont, F. J. (1982). The value of computed tomography in spinal trauma. *Spine* **7**, 417–31.

Dubois, P. J., Drayer, B. P., Sage, M., Osborne, D., and Heinz, E. R. (1981). Intramedullary penetance of metrizamide in the dog spinal cord. *Am. J. Neuroradiol.* **2**, 313–17.

Eismont, F. J., Clifford, S. Goldberg, M., and Green, B. (1984). Cervical sagittal spinal canal size in spine injury. *Spine* **9**, 663–6.

Epstein, N., Epstein, J. A., Benjamin, V., and Ransohoff, J. (1980). Traumatic myelopathy in patients with cervical spinal stenosis without fracture or dislocation—methods of diagnosis management and prognosis. *Spine* **5**, 489–96.

Evans, D. K. (1976). Anterior cervical subluxation. *J. Bone Joint Surg.* **58**(B), 318–21.

Farfan, H. F. (1984). The torsional injury of the lumbar spine. *Spine* **9**, 53.

Fielding, J. W. and Hawkins, R. J. (1977). Atlanto axial rotary fixation. Fixed rotary subluxation of the atlanto axial joint. *J. Bone Joint Surg.* **59**(A), 37–44.

Fielding, J. W., van Cochran, G. B., Lawsing, J. F., and Hohl, M. (1974). Tears of the transverse ligament of the atlas. *J. Bone Joint Surg.* **56**(A), 1683–91.

Fife, D. and Kraus, J. (1986). Anatomic location of spinal cord injury. Relationship to the cause of injury. *Spine* **11**, 2–5.

Forsyth, H. F. (1964). Extension injuries of the cervical spine. *J. Bone Joint Surg.* **46**(A), 1792–7.

Freeman, L. W. and Wright, T. W. (1953). Experimental observations of concussion and contusion of the spinal cord. *Ann. Surg.* **137**, 433–43.

Garret, J. W. and Braunstein, P. W. (1962). The seat belt syndrome. *J. Trauma* **2**, 220–38.

Gehweiler, J. A., Osborne, R. L., and Becker, R. F. (1980). *The radiology of vertebral trauma. Volume 16, Monographs in Clinical Radiology.* W. B. Saunders Company, Philadelphia, Pennsylvania.

Gehweiler, J. A., Clark, W. M., Schaaf, R. E., Powers, B., and Miller, M. D. (1979). Cervical spine trauma: the common combined conditions. *Radiology* **130**, 77–86.

Green, J. D., Harle, T. S., and Harris, J. H. (1981). Anterior subluxation of the cervical spine: hyperflexion sprain. *Am. J. Neuroradiol.* **2**, 243–50.

Grogono, G. J. S. (1954). Injuries of the atlas and axis. *J. Bone Joint Surg.* **36**(B), 297–410.

Grundy, D. J., McSweeney, T., and Jones, H. W. F. (1984). Cranial nerve palsies in cervical injuries. *Spine* **9**, 339–43.

Guttman, L. (1966). Traumatic paraplegia and tetraplegia in ankylosing spondylitis. *Paraplegia* **4**, 188–203.

Guttman, L. (ed.) (1973). In *Spinal cord injuries—comprehensive management and research* (2nd edition), pp. 47–8. Blackwell Scientific Publications, Oxford.

Hadden, W. A. and Gillespie, W. J. (1985). Multiple level injuries of the cervical spine. *Injury* **16**, 628–33.

Hampson, S., Coombs, R., and Hemingway, A. (1984). Fractures of the upper thoracic spine—an addition to the 'seat-belt' syndrome. *Br. J. Radiol.* **57**, 1033–4.

Handel, S. F. and Lee, Y. Y. (1981). Computed tomography of spinal fractures. *Radiol. Clin. N. Amer.* **19**, 68–89.

Handelberg, F., Bellemans, M. A., Opdecam, P., and Casteleyn, P. P. (1981). The use of computed tomography in the diagnosis of thoracolumbar injury. *J. Bone Joint Surg.* **63**(B), 336–41.

Harding, J. R., McCall, I. W., Park, W. M., and Jones, B. F. (1985). Fracture of the cervical spine in ankylosing spondylitis. *Br. J. Radiol.* **58**, 3–7.

Hardy, A. G. (1977). Cervical spinal cord injury without bony injury. *Paraplegia* **14**, 296–305.

Harris, J. H. (1978*a*). *The radiology of acute cervical spine trauma.* Williams and Wilkins, Baltimore, Maryland.

Harris, J. H. (1978*b*). Acute injuries of the spine. *Seminars in Roentgenol.* **13**, 53–68.

Harris, J. H., Edeiken-Monroe, B., and Kopaniky, D. R. (1986). A practical classification of acute cervical spine injuries. *Orth. Clin. N. Amer.* **17**, 15–30.

Haughton, S. (1866). On hanging, considered from a mechanical and physiological point of view. *London Edinburgh and Dublin Philos. Mag. and J. Sci.* **32**, 23–4.

Hohl, M. (1974). Soft tissue injuries to the neck in automobile accidents. *J. Bone Joint Surg.* **56**(A), 1675–82.

Holdsworth, F. W. (1963). Fractures, dislocations and fracture–dislocations of the spine. *J. Bone Joint Surg.* **45**(B), 6–20.

Howland, W. J., Curry, J. L., and Buffington, C. B. (1965). Fulcrum fractures of the lumbar spine. Transverse fracture induced by improperly placed seat belts. *J. Am. Med. Ass.* **193**, 240–1.

Jefferson, G. (1920). Fracture of the atlas vertebra. Report of four cases, and a review of those previously recorded. *Br. J. Surg.* **7**, 407–22.

Jefferson, G. (1927). Discussion on spinal injuries. *Proc. R. Soc. Med.* **21**, 625–48.

Kaufer, H. and Hayes, J. T. (1966). Lumbar fracture–dislocation. *J. Bone Joint Surg.* **48**(A), 712–30.

Kelly, R. P. and Whitesides, T. E. (1968). Treatment of lumbodorsal fracture–dislocations. *Ann. Surg.* **167**, 705–17.

Kilcoyne, R. F., Mack, L. A., King, H. A., Ratcliffe, S. S., and Loop, J. W. (1983). Thoracolumbar spine injuries associated with vertical plunges: reappraisal with computed tomography. *Radiology* **146**, 137–40.

Kingma, L. M. (1986). Radiological analysis in case of trauma to the lumbar and thoracic spine. In *Neuroradiology 1985/1986. Proceedings of the XIII Congress of the European Society of Neuroradiology, Amsterdam, 11–15 September 1985* (ed. J. Valk), p. 350. Excerpta Medica Foundation, Amsterdam.

Levine, A. M. and Edwards, C. C. (1986). Treatment of injuries of the C1–2 complex. *Orth. Clin. N. Amer.* **17**, 31–44.

Lewis, J. and McKibbin, B. (1974). The treatment of unstable fracture–dislocations of the thoracolumbar spine accompanied by paraplegia. *J. Bone Joint Surg.* **56**(B), 603–12.

Lindahl, S., Willen, J., and Irstam, L. (1983). Computed tomography of bone fragments in the spinal canal. An experimental study. *Spine* **8**, 181–6.

Lindahl, S., Willen, J., Nordwall, A., and Irstam, L. (1983). The crush–cleavage fracture. A 'new' thoracolumbar unstable fracture. *Spine* **8**, 559–69.

Louis, R. (1977). Symposium instables du rachis. Les théories de l'instabilité. *Rev. Chir. Orthop.* **63**, 423–5.

MacGibbon, B. and Farfan, H. F. (1979). A radiologic survey of various configurations of the lumbar spine. *Spine* **5**, 258–66.

MacNab, I. (1971). The 'whiplash syndrome'. *Orth. Clin. N. Amer.* **2**, 389–403.

Malgaigne, J. F. (1855). *Traité des fractures et des luxations, T. II.* Baillière, Paris.

McAfee, P. C., Yuan, H. A., and Lasda, N. A. (1982). The unstable burst fracture. *Spine* **7**, 365–73.

McKenzie, J. A. and Williams, J. F. (1971). The dynamic behaviour of the head and cervical spine during 'whiplash'. *J. Biomech.* **4**, 477–90.

Merskey, H. (1984). Psychiatry and the cervical pain syndrome. *Can. Med. Ass. J.* **130**, 1119–21.

Miller, H. (1961). Accident neurosis. *Br. Med. J.* **1**, 919–25, 992–8.

Miller, M. D., Gehweiler, J. A., Martinez, S., Charlton, O. P., and Daffner, R. H. (1978). Significant new observations on cervical spine trauma. *Am. J. Roentgenol.* **130**, 659–63.

Mudge, K., van Dolson, L., and Lake, A. S. (1984). Progressive cystic degeneration of the spinal cord following spinal cord injury. *Spine* **9**, 253–5.

Nachemson, A. (1960). Fracture of the odontoid process of the axis. A clinical study based on 26 cases. *Acta Orthopaed. Scand.* **29**, 185–217.

Newman, R. J. (1984). Chest wall injuries and the seat belt syndrome. *Injury* **16**, 110–13.

Nicholas, J. A., Wilson, P. D., and Freiberger, R. H. (1960). Pathological fractures of the spine: etiology and diagnosis. *J. Bone Joint Surg.* **42**(A), 127–37.

Nicoll, E. A. (1949). Fractures of the dorso-lumbar spine. *J. Bone Joint Surg.* **31**(B), 376–94.

Panjabi, M. M., Hausfield, H. N., and White, A. A. (1981). A biomechanical study of the ligamentous stability of the thoracic spine in man. *Acta Orthopaed. Scand.* **52**, 315–26.

Paradis, G. R. and Jones, J. M. (1973). Post-traumatic atlanto–axial instability: the fate of the odontoid process in 46 cases. *J. Trauma* **13**, 359–67.

Pelker, R. R. and Dorfman, G. S. (1986). Fracture of the axis associated with vertebral artery injury. *Spine* **11**, 621–3.

Penning, L. (1968). *Functional pathology of the cervical spine*, p. 120. Excerpta Medica Foundation, Amsterdam.

Quencer, R. M., Green, B. A., and Eismont, F. J. (1983). Posttraumatic spinal cord cysts: clinical features and characterization with metrizamide computed tomography. *Radiology* **146**, 415–23.

Riggins, R. S. and Kraus, J. F. (1977). The risk of neurological damage with fractures of the vertebrae. *J. Trauma* **17**, 126–33.

Roaf, R. (1960). A study of the mechanics of spinal injury. *J. Bone Joint Surg.* **42**(B), 810–23.

Roaf, R. (1963). Lateral flexion injuries of the cervical spine. *J. Bone Joint Surg.* **45**(B), 36–8.

Roaf, R. (1972). International classification of spinal injuries. *Paraplegia* **10**, 78–84.

Roberson, J. R. and Whitesides, T. E. Jr. (1985). Surgical reconstruction of late post-traumatic thoracolumbar kyphosis. *Spine* **10**, 307–12.

Roberts, A. and Wickstrom, J. (1973). Prognosis of odontoid fractures. *Acta Orthopaed. Scand.* **44**, 21–30.

Roberts, J., and Curtis, P. H. (1970). Stability of the thoracic and lumbar spine in traumatic paraplegia following fracture or fracture–dislocation. *J. Bone Joint Surg.* **52**(A), 1115–30.

Rogers, L. F. (1981). Fractures and dislocations of the spine. In *Radiology of spinal cord injury* (ed. L. Calenoff) Chapter 4, pp. 85–129. C. V. Mosby Company, St Louis, Missouri.

Rogers, L. G. (1971). The roentgenographic appearance of transverse or chance fracture of the spine: the seat belt fracture. *Am. J. Roentgenol.* **111**, 844–9.

Schaaf, R. E., Gehweiler, J. A., Miller, D., and Powers, B. (1978). Lateral hyperflexion injuries of the cervical spine. *Skeletal Radiol.* **3**, 73–8.

Schatzker, J., Rorabeck, C. H., and Waddell, J. P. (1971). Fractures of the dens (odontoid process). An analysis of 37 cases. *J. Bone Joint Surg.* **53**(B), 392–405.

Scher, A. T. (1979). Anterior cervical subluxation: an unstable position. *Am. J. Roentgenol.* **133**, 275–80.

Scher, A. T. (1982). 'Tear drop' fractures of the cervical spine—radiological features. *S. Afr. Med. J.* **61**, 355–6.

Schneider, R. C. and Kahn, E. A. (1956). Chronic neurological sequelae of acute trauma to the spina and spinal cord. Part I. The significance of the acute-flexion or 'tear-drop' fracture–dislocation of the cervical spine. *J. Bone Joint Surg.* **38**(A), 985–97.

Schneider, R. C., Livingston, K. E., Cave, A. J. E., and Hamilton, G. (1965). Hangman's fracture of the cervical spine. *J. Neurosurg.* **22**, 141–54.

Seibert, C. E., Dreisbach, J. N., Swanson, W. B., Edgar, R. E., Williams, P., and Hahn, H. (1981). Progressive posttraumatic cystic myelopathy: neuroradiologic evaluation. *Am. J. Neuroradiol.* **2**, 115–19.

Smith, W. S. and Kaufer, H. (1969). Patterns and mechanisms of lumbar injuries associated with lap seat belts. *J. Bone Joint Surg.* **51**(A), 239–54.

Spence, K. F., Decker, S., and Sell, K. W. (1970). Bursting atlantal fracture associated with rupture of the transverse ligament. *J. Bone Joint Surg.* **52**(A), 543–9.

Stevens, J. M., Olney, J. M., and Kendall, B. E. (1985). Post-traumatic cystic and non cystic myelopathy. *Neuroradiology* **27**, 48–56.

Streiter, M. L. and Chambers, A. A. (1984). Metrizamide examination of traumatic lumbar nerve root meningocele. *Spine* **9**, 77–8.

Sullivan, R. C., Bruwer, A. J., and Harris, L. (1958). Upper mobility of the cervical spine in children. A pitfall in the diagnosis of cervical dislocation. *Am. J. Surg.* **95**, 636–40.

Swain, A., Grundy, D., and Russell, J. (1986). ABV of spinal cord injury. Early management and complications. *Br. Med. J.* **292**, 44–7.

Taylor, A. R. and Blackwood, W. (1948). Paraplegia in hyperextension cervical injuries with normal radiographic appearances. *J. Bone Joint Surg.* **30**(B), 245–8.

Wang, G. J., Mabie, K. N., Whitehill, R., and Stamp, W. G. (1984). The nonsurgical management of odontoid fractures in adults. *Spine* **9**, 229–30.

Webb, J. K., Broughton, R. B. K., McSweeney, T., and Park, W. M. (1976). Hidden flexion injury of the cervical spine. *J. Bone Joint Surg.* **58**(B), 322–7.

Weir, D. C. (1975). Radiographic signs of cervical injury. *Clin. Orthop.* **109**, 9–17.

Weller, R. O. (1984). *Colon atlas of neuropathology.* Harvey Miller, London.

Weller, R. O., Swash, M., McLellan, D. L., and Scholtz, C. L. (1983). *Clinical neuropathology.* Springer-Verlag, Berlin.

White, A. A. and Panjabi, M. M. (1978). *Clinical biomechanics of the spine*, pp. 211–14. J. B. Lippincott, Philadelphia, Pennsylvania.

White, A. A., Southwick, W. O., and Panjabi, M. M. (1976). Clinical instability in the lower cervical spine. *Spine* **1**, 15–27.

White, A. A., Johnson, R. M., Panjabi, M. M., and Southwick, W. O. (1975). Biomechanical analysis of clinical stability of the cervical spine. *Clin. Orthop.* **109**, 85–96.

White, P. R., Newberg, A., and Seligson, D. (1980). Computerized tomographic assessment of the traumatized dorsolumbar spine before and after Harrington instrumentation. *Clin. Orthoped. Related Res.* **146**, 150–6.

Whitesides, T. E. (1977). Traumatic kyphosis of the thoracolumbar spine. *Clin. Orthop.* **128**, 78–92.

Whitley, J. E. and Forsyth, H. F. (1960). The classification of cervical spine injuries. *Am. J. Roentgenol.* **83**, 633–44.

Willen, J., Lindahl, S., Irstam, L., Aldman, B., and Nordwall, A. (1984). The thoracolumbar crush fracture. An experimental study on instant axial dynamic loading: the resulting fracture type and its stability. *Spine* **9**, 624–31.

Woodford, M. and Walford, G. (1985). Radiography in modern spinal treatment centre. *Radiography* **51**, 269–73.

Wood-Jones, F. (1913). The ideal lesion produced by judicial hanging. *Lancet* **i**, 53.

II

Clinical assessment of spinal cord injury
(theoretical and practical)

4

Clinical evaluation and pathophysiology of the spinal cord in the chronic stage
L. S. Illis

Introduction

The clinical picture in the chronic stage of spinal cord dysfunction is one of the more common problems of neurological rehabilitation. The management of such a patient is conventionally regarded as the management or treatment of the complications which arise as a result of dysfunction. However, if the clinical picture is seen, not in isolation, but as part of the gradual evolution from the acute phase of the lesion then the clinical picture becomes less that of a patient who is a burden to the community and more that of a person who is, at least potentially, amenable to treatment, and so forms a link between academic research based on plasticity of the central nervous system (CNS) and clinical neurology. Seen in this light, the neurological examination is not an academic exercise in localization but an essential part of the interpretation of the mechanism of recovery. Together with neurophysiological measurements and investigation, it should be possible to delineate the neurological disturbance in structural and functional terms and to attempt to alleviate distress and improve function on a scientific rather than on an *ad hoc* basis.

Prognostic information regarding the extent of the spinal lesion, whether the lesion is clinically complete or incomplete, the relative involvement of motor or sensory components, and the presence and severity of associated injuries can only be properly assessed by repeated and carefully documented examination.

Though the clinical neurological examination gives an accurate description of the upper level of the lesion, it does not give a representation of the extent of the damaged spinal cord. It is an examination of integrated function. However, in the chronic stage of spinal cord injury the CNS has disintegrated, as it were, below the level of the lesion and what becomes more relevant is not the loss of integrated function (which can really only describe the site of the lesion) but what is preserved, even though it is disintegrated. In effect, a new type of neurology is needed—the neurology of disintegration.

For example, a simple 'upper motor neuron' or 'pyramidal' disturbance probably never occurs. Rather, there are at least three components; i.e. volitional mechanisms, postural mechanisms, and equilibrium mechanisms. The relative disturbance of these three components cannot be assessed unless the patient is examined not only lying down, but also sitting and (where possible) standing because of the postural mechanisms. Even if volitional mechanisms are totally absent when the patient is lying down, if postural control is present on standing and a withdrawal reflex can be elicited then this is a different degree of disintegration than in a patient where postural control is absent. The essential degree of improvement, for example, with stimulation techniques, is quite different in the two cases (Dimitrijevic, Illis, Nakajima, Sharkey, and Sherwood 1986a; Dimitrijevic, Dimitrijevic, Illis, Nakajima, Sharkey, and Sherwood 1986b). Clearly, suprasegmental influences are necessary, and probably some residual function of posterior columns is also necessary for spinal segmental mechanisms to be susceptible to modulation (Saade, Tabet, Atweh, and Jabbur 1984). A lesion in the CNS does not comprise a simple matter of destruction of certain pathways which were present in the normal or intact CNS, in that it is not just the removal (by destruction) of various unrelated components but a gradual alteration of the CNS that has occurred. To understand this it is necessary to start with the acute phase and follow the changes which occur.

The fact that the chronic state is different from the acute state must indicate that one of two events has occurred: either the lesion has progressed or regressed in order to produce the chronic state, or the CNS has altered as a result of the lesion. If we take the simplest form of disturbance, namely an acute injury, then this can be examined in more detail. A single insult will be accompanied by changes at the normal–abnormal interface, producing transient changes in cells, metabolic disturbances, and similar changes. This may be responsible for short-term changes in the clinical state and may be of significance in early management, but it is of little importance in the context of the chronic state. Although structural changes directly related to the lesion may occur and are of great clinical importance (e.g. cyst formation) they are usually clinically recognizable.

This leaves the evolution of the neurological picture from the acute to the chronic state, which can only be due to an evolution of the CNS reaction, and it is this which needs to be analysed in order to put the clinical evaluation of spinal cord dysfunction in the chronic stage into a neurological perspective. The changing clinical picture is, in effect, a reflection of recovery of function—however incomplete or abortive—and can be seen as a change in anatomy, physiology, and pharmacology, using spinal shock as an example of recovery. At present, nothing can be done about the lesion in terms of recovery, at least in the chronic state, except to teach the patient to live with his disability and to prevent or treat complications. More detailed analysis of the altered intact CNS may lead to practical methods to further recovery. In addition, a study of pharmacological changes as well as the anatomy and physiology of the damaged nervous system throws light on the true nature of the neurological deficit seen after a lesion. Unless this is examined we cannot understand, let alone influence, the chronic state of damage.

The nature of recovery

In its simplest form, there are three main bodies of opinion regarding the nature of recovery: Von Monakow's (1914) concept of 'diaschisis' (often used as an explanation, although it is purely a description); sprouting and unmasking—i.e. structural–physiological changes; and alteration of synaptic effectiveness, the area in which investigation of the nature of recovery and possible therapy most overlap. The three groups are not mutually exclusive. Von Monakow's concept of diaschisis was suggested as the explanation for long-delayed spontaneous recovery after a lesion in the CNS. Von Monakow suggested that if a part of the CNS is destroyed, a distant part with which it was in neuronal contact may stop functioning. After some period of time the 'depressed' area recovers its ability to function. He was unable to explain the mechanism of recovery, but it is clear that recovery of function would be due to removal of 'depression' with no reorganization or regeneration of the CNS taking place. Recent experimental work suggests a possible pharmacological basis for diaschisis.

Pharmacological changes

There are probably at least three types of communication between nerve cells: point-to-point or fast transmission (conventional neurotransmission); neuromodulation, involving the direct alteration of post-synaptic conductance, probably by altering transmitter release or by altering post-synaptic responsiveness; and neurohormonal communication, in which there is no anatomical contiguity via synaptic contacts. In neurohormonal communication there are long-lasting changes in target cells following stimulation of specific peptidergic neurons. In contrast to fast transmission, monoamine release is associated with diffuse neural pathways, with nerve cells localized in the brain stem and with diffuse ramifications to wide terminal fields, so that very large numbers of target cells may be affected. These anatomical factors suggest that such a system is modulatory and this is the generally accepted view (Iversen 1982). These transmitter systems include the catecholamines: adrenalin, noradrenalin, and dopamine. In the context of possible mechanisms in the chronic stage of spinal cord dysfunction an important system is the noradrenalin system which ramifies to virtually all areas of the CNS, originating mostly from brain-stem nuclei (*locus coeruleus*) and innervating the cerebral cortex, cerebellum, hippocampus, hypothalamus, and spinal cord. Brailowski (1980) has summarized the effect of many pharmacological agents and the factors which may modify their use. Ever since Wolfe's (1940) report, scientists have used drugs to influence recovery following the experimental application of lesions to the CNS. Kennard and colleagues (Ward and Kennard 1942; Watson and Kennard 1945) used cholinergic agents in cortical ablation experiments in monkeys and demonstrated an increase in the number of animals recovering. Acetyl-choline in intrinsic CNS pathways is probably modulatory in effect, as opposed to its fast transmitter action in spinal cord and cranial motor neurons. Russian workers headed by Luria (Luria, Naydin, Tsvetkova, and Vinarskaya 1969) have reported beneficial results, but these have hardly been followed up in the UK.

Endogenous opiates may interact with other physiological systems, particularly in the autonomic nervous system. Opioids are co-stored with catecholamines in sympathetic ganglia and adrenal medulla, and in some species are released with catecholamines and controlled in the same way as catecholamines. The opioid antagonist, naloxone, has been used as a blocking agent to investigate endogenous opiate function and has been used for a variety of neurological disorders. In spinal injury (cat) naloxone has been reported to decrease neurological impairment and to increase spinal cord blood-flow (Faden, Jacobs, Mougey, and Holaday 1981; Young, Flamm, Demopoulos, Tomasula, and Decrescito 1981). Improvement has also been reported in the treatment of stroke (Baskin and Hosobuchi 1981), experimental hemiplegia (Baskin, Keick, and Hosobuchi 1982), and sub-acute necrotizing encephalopathy (Brandt, Terenius, Jacobson, Klinken, Nordus, Brandt, Blegvad, and Yssing 1980). Thus pharmacological reversal of neurological deficit is a possibility and suggests that some part of the neurological picture is produced by pharmacological methods acting via a neuromodulatory system.

The use of catecholamine agonists and antagonists is of more interest, as indicated earlier. Several studies have reported improved behavioural changes in the case of CNS lesions (rat) with the use of d-amphetamine and methamphetamine (Krieckhause 1965; Braun, Meyer, and Meyer 1966; Cole, Sullins, and Isaac 1967; Bennett, Rosenzweig, and Wu 1973). *Locus coeruleus* stimulation enhances monosynaptic reflexes and facilitates afferent impulse transmission. The potentiating effect can be abolished by systemic phenoxybenzamine (Fung and Barnes 1984).

Experimentally, the catecholamine system shows some potential in the treatment of neurological deficit and raises many questions as to the true nature of such deficit. A study of this system indicates perhaps the first correct explanation of the descriptive concept of diaschisis. For example, unilateral sensory–motor cortex ablation (rat and cat) produces a specific deficit which can be altered pharmacologically in a profound way (Feeney,

Gonzales, and Law 1982). A single injection of amphetamine (12 mg/kg) produces a marked recovery of neurological deficit, providing the animal is kept mobile. The reversal of neurological deficit does not occur if the animal is confined, so presumably sensory feedback is essential for the recovery process. The recovery can be retarded or reversed by catecholamine antagonists such as phenoxybenzamine (Feeney, Hovda, and Salo 1983) and can be blocked by haloperidol when given early after the injury. This work has not yet been followed by therapeutic trials in man, and the concept has not been applied to spinal injury. The reversal of deficit and the time scale is such that the recovery cannot be due to the direct pharmacological effect on the original, primary, damage, but must be due to an effect on disturbed function remote from the lesion. This suggests that the original insult to the CNS produces its effects in two distinct but interrelated ways. On the one hand, there is obvious damage to the CNS, with some recovery at the interface with normal tissue due to reversible damage and circulatory and metabolic effects. This produces the 'fixed' lesion and fixed deficit of classical and conventional neurology, with the clinical syndrome relating directly to the damaged area of CNS on a simple cause-and-effect basis, narrowing the search for the basis of recovery to the site of the lesion and, not surprisingly, producing decades of somewhat abortive research. On the other hand, there is an undoubted, and, as yet, unexplained process of recovery which cannot relate to the original insult and must be due to some functional abnormality of areas of CNS which are necessary for the expression or performance of the behaviour lost as a result of the lesion. The noradrenalin system, with its widespread ramifications to a broad target area and its modulatory effect, is a potential candidate for the pharmacological basis for diaschisis (alone or with the neuropeptide system). We are dealing, however, with an abnormal nervous system, one which has altered structurally and physiologically, and the pharmacological changes, therefore, must be seen in this context.

This leads to the second main opinion as to the nature of recovery of function. This is a corollary of plasticity of the nervous system, so that recovery is consequent upon a reorganization of the nervous system through the sprouting of new synapses or the unmasking of existing synapses.

Sprouting of intact fibres
There are more nerve cells present during embryonic life than in the mature nervous system and it appears that developing neurons must make synaptic contact with appropriate target cells in order to survive (Purves and Nja 1978). One suggested mechanism for the cessation of nerve growth when appropriate target cells are contacted is contact inhibition; i.e. neurons carry a finite number of terminals and if a proportion of these degenerate due to axonal injury then the uninjured axons sprout to fill the local sites available

and consequently expand the synaptic field of the uninjured axons. This has implications in terms of altered physiology since the relative contributions of afferent axons are now altered. Removal or alteration of contact inhibition (possibly by degeneration metabolites) will result in continuous remodelling in the intact CNS, or regeneration and reorganization of the damaged CNS. Collateral sprouting is recognized as a widespread phenomenon and has been demonstrated in peripheral, central, and autonomic systems (Edds 1953; Murray and Thompson 1957; Liu and Chambers 1958; Illis 1967b; Raisman 1969; Lynch, Deadwyler, and Cotman 1973). However, although the fact of sprouting is unquestioned, its significance remains uncertain. Is it regeneration and an attempt to restore normal function, or is it a random response resulting in a haphazard and inappropriate connectivity, or is it a combination of both? Is it initiated by the lesion or is it a natural and continuing process in the normal CNS? Examination of protein synthesis after spinal cord lesions (Bernstein and Wells 1980) and of axoplasmic transport (Lasek and Hoffmann 1976) gives evidence in favour of the idea that nerve cells are in a state of continual replacement. When a lesion occurs, a response in terms of an increase in the level of amino-acids and in protein metabolism is seen within hours. However, sprouting alone cannot explain short-term changes in the nervous system and, furthermore, the alteration of the relative contribution of terminals to the target site will produce an abnormal connectivity which, though contributing to the altered clinical state, is not necessarily contributing to recovery.

Unmasking of existing synapses

Wall and his colleagues have opened up a new and important field of plasticity in the CNS; i.e. the unmasking of synapses. Merrill and Wall (1978) have reviewed a series of experiments in *adult* animals (rat and cat) which demonstrates that when cells are deafferented, and therefore lose their normal input, they begin to respond to new inputs, i.e. inputs which in the intact animal produce no response. This switching from normal to new input may occur immediately and can, by cold reversible block, be switched back and forth. This immediate switching was found in dorsal column nuclei, but for the majority of cells the change takes days or weeks to complete. Unmasking is not only seen following degeneration, since peripheral nerve section by itself produces no signs of central degeneration and therefore no sprouting to establish new connections, but there is also a central alteration of receptive fields and this is probably due to the change in afferent bombardment (Devor and Wall 1976). The fact that the altering of peripheral stimulation alters receptive fields in the CNS has major implications for the rehabilitation of neurological deficit in man. For example, in both sprouting and unmasking, previously unused or little-used pathways now take on a more significant role. For example, Goldberger and Murray

(1974) have demonstrated the occurrence of reflex activity unmasked by chronic deafferentation; i.e. reflex activity which could not be elicited in the intact animal could be elicited (unmasked) following the occurrence of a lesion in the intact nervous system.

Faganel and Dimitrijevic (1982) studied the effect on ankle jerks, in patients with clinically complete chronic spinal cord injury, after conditioning with noxious electrical stimuli applied to thoracic, lumbar, and sacral dermatomes. Painful cutaneous stimulation applied to the skin produces a reflex contraction and withdrawal. The organization includes several segments activated as functional units. In man, in the case of a complete transverse lesion the induced excitation spreads to adjacent and more distant cranial and caudal segments both ipsilaterally and contralaterally (Sherrington 1910). This spread of reflexes, not seen in the normal or intact spinal cord, reflects an altered anatomy and physiology. The alteration in anatomy can be explained in terms of either sprouting and/or unmasking, since the existence of crossed monosynaptic afferents and the spread over adjacent and distant segments has been demonstrated anatomically (Illis 1967*a*, 1973*a*) and the alteration in physiology is on the basis of such anatomy; i.e. in the intact animal the anatomical substrate is present but is 'masked' by a dominant system of connections. One effect of the lesion is to alter this dominance by sprouting and unmasking of less dominant, but pre-existing, anatomical systems.

Alteration in synaptic transmission or effectiveness
Denervation supersensitivity, the increase in sensitivity of synapses or of post-synaptic elements following axonal damage, is well documented in muscle and autonomic ganglia. Possibly a similar mechanisms exists in the CNS, but the evidence for this is poor.

Alteration of signal traffic may change synaptic transmission and effectiveness. The best known example of this is post-tetanic potentiation. As already indicated, an intriguing series of experiments by Devor and Wall (1976; and see Devor 1982) is of the greatest interest in rehabilitation. In these experiments, deafferentation of dorsal horn cells was carried out by cutting the nerves distal to the dorsal root ganglion so that no degeneration could occur in the cord. Following this the cells became responsive to distant inputs. The most likely explanation of this would be an alteration in signal traffic, in terms of timing or of pattern.

Repetitive stimulation may alter appropriate synapses anatomically (Illis 1969). There is abundant evidence for environmental stimulation and repetitive electrical stimulation producing structural, physiological, and behavioural changes (see Illis 1982) in animals and in man. Spinal cord stimulation, a technique which arose directly from the 'gate-control' hypothesis of Melzack and Wall (1965), produces clinical and physiological changes (Cook and Weinstein 1973; Illis, Sedgwick, and Taylor 1980; Sedgwick, Illis, Tallis, Thornton, Abraham, El-Negamy, Docherty, Soar, Spencer, and Taylor 1980; Illis, Read, Sedgwick, and Tallis 1983; Sherwood 1985; and see Table 4.1). The mode of action of spinal cord stimulation is unknown, but in some way the pattern of signal traffic is altered, perhaps altering central inhibition and perhaps unmasking synapses.

Table 4.1. Reported physiological changes with spinal cord stimulation

Author(s)	Animal	Effect
Frolich and Sherrington 1902	Cat Dog Monkey	Inhibition of muscular rigidity
Larson, Sances, Riegel, Meyer, Dallman, and Swiontek 1974	Man Monkey	Changes in evoked responses
Bantli, Bloedel, and Thienprasit 1975	Monkey	Evoked responses in thalamic nuclei
Foreman, Beall, Applebaum, Coulter, and Willis 1976	Primate	Changes in spinothalamic tract neurons; post-synaptic inhibition
Siegfried, Krainick, Haas, Adoriani, Meyer, and Thoden 1978	Cat	Inhibition of monosynaptic reflex activity
	Man	Decrease in tonic stretch reflex; changes in H-reflex
Saade *et al.* 1984	Cat	Modulation of spinal segmental mechanisms: primary afferent depolarization; depression of polysynaptic reflexes; facilitation of monosynaptic reflexes
Wiesendanger, Chapman, Marini, and Schorderet 1985	Cat Monkey	Depression of stretch reflexes (or facilitation, depending on electrode position)
Illis, Sedgwick, Oygar, and Awadalla 1976	Man	Changes in H-reflex
Thoden, Krainick, Strassburg, and Zimmerman 1977	Man	Changes in H-reflex
Abbate, Cook, and Attalah 1977	Man	Improvement in urodynamic studies
Sedgwick *et al.* 1980	Man	Changes in cervical and brain-stem evoked responses
Feeney and Gold 1980	Man	Changes in H-reflex
Illis *et al.* 1980	Man	Improvement in urodynamic studies

Table 4.1 *continued*

Author(s)	Animal	Effect
Read, Matthews, and Higson 1980	Man	Improvement in urodynamic studies
Hawkes, Wyke, Desmond, Bultitude, and Kanegaonkar 1980	Man	Improvement in urodynamic studies
Meglio, Cioni, d'Amico, Ronzoni, and Rossi 1980	Man	Improvement in urodynamic studies
Dimitrijevic and Sherwood 1980	Man	Alteration of stretch reflexes
Berg, Bergmann, Hovdal, Hunstead, Johnson, Levin, and Sjaastad 1982	Man	Improvement in urodynamic studies
Shimoji, Shimizu, Maruyama, Matsuki, Kurirayashi, and Fujioka 1982	Man	Pre-synaptic inhibition; facilitation of primary afferent depolarization
Laitinen and Fugl-Meyer 1981	Man	Improvement in spasticity (isokinetic dynamometry measurement)
Scerrati, Onofrj, and Pola 1982	Man	Changes in H-reflex
Tallis, Illis, Sedgwick, Hardwidge, and Garfield 1983; Tallis, Illis, Sedgwick, Hardwidge, and Kennedy 1983	Man	Increase in skin and muscle blood-flow
Read, James, and Shaldon 1985	Man	Improvement in urodynamic studies
Nakamura and Tsubokawa 1985	Man	Changes in H-reflex; reduction in spasticity
Barolat-Romana, Myklebust, Hemmy, Myklebust, and Wenninger 1985	Man	Inhibition of spasms: EMG changes
Gottleib, Myklebust, Stefoski, Groth, Kroin, and Penn 1985	Man	Alteration of stretch reflexes; changes in joint compliance

Spinal shock

Wall (1980) has pointed out that:

... quite serious text books state that the early areflexia of paraplegia is caused by spinal shock. 'Spinal shock' is not a cause of anything, it is a circular confusing description of the state of areflexia. Furthermore, at a later stage, the paraplegic is not recovering from spinal shock, he is in fact recovering.

The term 'spinal shock', introduced by Marshall Hall in 1850, was first described 100 years earlier by Whytt in 1750 (both quoted in Sherrington 1910). The term is used to signify the effect of sudden injury or transection of the spinal cord. It is characterized by sensory and motor paralysis and then, later, by the gradual recovery of reflexes. Nobody has ever given a convincing explanation of the recovery of reflexes following their complete abolition. Spinal shock occurs at any level of transection below the mid-*pons*. Above this level a transection produced decerebrate rigidity. Following transection below the *pons*, nerve cells in the spinal cord are suddenly cut off from all descending influences and then undergo a dramatic change in their capacity to react. The relative importance of different pathways in producing spinal shock is poorly known and there is no uniformity of opinion. However, in lower mammals the important descending influences appear to be reticulospinal and vestibulospinal, whereas in higher animals, including man, corticospinal connections are probably more important. The intensity and the duration of spinal shock increases as the vertebrate scale is ascended and probably parallels the degree of development of suprasegmental structures. Guttman (1976) described intractable spasticity in all four limbs of a patient with a cervical lesion. Intrathecal block with alcohol in this patient blocked the thoracolumbar section of the spinal cord and this procedure resulted in decreased spasticity in the arms. This clinical report emphasizes the importance of the *afferent* system in the production of the clinical picture (cf. the importance of sensory feedback in the reversal of neurological deficit following the injection of amphetamine, and the importance of altered signal traffic).

It is usually assumed that the importance of pathways in producing spinal shock is a reflection of their specificity. It may equally be argued, however, that the importance lies in their relative contribution to the total spinal cord input. The fact that the depression of spinal shock is much more severe when transection is below the *pons* does not necessarily mean that there is some particularly important aborally directed control arising in the *pontine* nuclei; it could mean that the difference in transection above and below the *pons* is due to the relative contribution to the descending pathways made by different levels of the neuraxis. The greater the number of descending pathways interrupted, the more likely that shock will ensue, because the more severely disorganized will be the target area upon which these descending pathways converge. The higher contribution of inputs from a particular centre does, of course, confer some specificity, but it is not necessary to assign the capacity for the production of spinal shock onto the centre or its pathways.

There has not been a great deal of progress in the understanding of spinal shock in the last twenty-five years. Denny-Brown (1960), after reviewing the features of spinal shock, and theories as to its causation and recovery, concluded that the nature of spinal shock is obscure. The phenomenon of miniature end-plate potentials, produced by small quanta of transmitter impulses (Katz and Miledi 1962), could perhaps explain how the loss of a potent transmitter pathway could result in depression of all excitation of the motor or internuncial neuron, but it does not explain why reflexes recover in an altered form. Most theories regarding the first phase of spinal shock (depression of reflexes) include the withdrawal of impulses from descending tracts, but no explanation of spinal shock can be acceptable unless it includes the second phase—that of the return of reflexes in an altered form.

Denny-Brown (1951) described the peculiar lack of response which occurs following leucotomy in man and animals and which lasts for 3–21 days: 'the disequilibrium of cortical mechanisms following abolition of the effect of one of them . . . preceded by a period of depression of all the others, which we think is the explanation of the "diaschisis" of Monakow'. This illustration (hardly an explanation) of 'diaschisis' is not unlike that of spinal shock, and serves to remind us that the concept of spinal shock is not in fact limited to the spinal cord, and also reminds us of how little our knowledge has advanced since the early work of von Monakow.

There are two further features of spinal shock which are of particular interest. First, it has been commonly assumed that the effects of transection occur in an aboral direction only. This is not so, since headward of the transection there is a change in the reflex response (Creed, Denny-Brown, Eccles, Liddell, and Sherrington 1932). Second, there are the experiments of Teasdell and Stavraky (1953) to consider, in which one limb of a cat was deafferented by section of the posterior roots; electrical stimulation of the basis pedunculi produced no response in the corresponding limb, but, 5–47 days later, responses were evoked more readily in the denervated than in the normal limbs. This is a similar but reversed state of affairs to that pertaining to spinal shock and has particular reference to some of the work described.

Sherrington (1910) summarized spinal shock as a withdrawal of synaptic transmission. Quantitative studies of the motor neuron surface (Armstrong, Richardson, and Young 1956; Wyckoff and Young 1956; Armstrong and Young 1957; Aitken and Bridger 1961; Illis 1964a, 1964b, 1973a, 1973b) have indicated the complexity of the 'synaptic zone', and it has been suggested that this zone may be profoundly altered by denervation of a critical number of *boutons termineaux*, with the consequences not only of degeneration of those *boutons* which were deafferented but also of disorganization of the whole synaptic zone (Illis 1963, 1967b). This would result in the depression of activity of nerve cells. With time, *boutons* with severed afferents would degenerate and those with intact fibres would reorganize.

The reorganization would be produced by sprouting (Liu and Chambers 1958; Illis 1967*a*, 1973*a*, 1973*b*; Raisman 1969;) and by unmasking (Merrill and Wall 1978). The result of this reorganization in terms of reflexes would be that any reflex-eliciting stimulus would now produce a response which, though more or less specific at any one time, would be changing along with the changes taking place in the reorganization of the network. This theoretical concept has had some confirmation in the work of Goldberger and Murray (1974) and Faganel and Dimitrijevic (1982), as reported above. The abnormal pattern, rather than the abnormal reflex, has been little studied in chronic lesions of the CNS, and interpretation of the pathological spread of reflexes in terms of reorganization is scant. However, recent work, on the effect of spinal cord stimulation in patients with chronic spinal cord injury, has tended, again, to confirm the theoretical concept (Dimitrijevic *et al.* 1986*a*, 1986*b*).

Spinal reflexes, more or less unrestrained by supraspinal inhibitory (co-ordinating?) influences, return in a much altered form characterized by hyperreflexia, rigidity and spasticity, and the presence of abnormal reflexes. At what point has the abnormal reflex activity returned? Is there, indeed, any end-point, or does the reflex activity continue to alter depending upon the effect of the *afferent* system? Does the clinical picture truly reflect the destruction caused by the lesion, or is it more accurately a reflection of the altered nervous system reacting to afferent stimuli (see Fig. 4.1)?

Return of reflexes

Once the first phase of spinal shock subsides, natural peripheral stimulation elicits a central response which is now free from descending influences and results in a complex reflex automatism characterized by hyperreflexia and hypertonia. The pattern of reflex return and the sequence is complicated.

The anatomical changes, summarized above, show a progressive change, which may begin within twenty-four hours of injury. In the cat, reflexes below the lesion are obtainable after about one hour. But not all reflexes are so readily obtainable, and over long periods of time reflexes are still changing in their nature. They come back in the order: knee jerk, crossed extensor, and extensor thrust; the last to appear is the scratch reflex. Sherrington (1910) observed that the scratch reflex six weeks after cord section was still irregular, feeble, and easily fatigued, and this was confirmed by Fulton and Sherrington (1932).

Recent techniques in clinical neurophysiology may impose a new interpretation on findings in man. For example, Dimitrijevic, Dimitrijevic, Faganel, and Sherwood (1984) have demonstrated suprasegmentally-induced responses in forty of fifty-eight paralysed patients. The demonstration of residual and potentially functioning long descending fibres through the severely damaged spinal cord (clinically paralysed patients) must con-

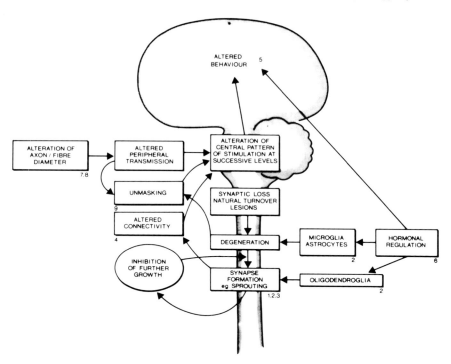

Fig. 4.1. Diagram to summarize the alterations in the synaptic zone and in behaviour with CNS damage. (a) Liu and Chambers 1958; (2) Illis 1973*a*, 1973*b*; (3) Cotman, Nieto-Sampedro, and Harris 1981; (4) Illis 1967*b*, 1982; (5) Marshall 1984; (6) Scheff, Anderson, and Dekosky 1984; (7) Jacobs and Love 1985; (8) Schaumberg, Spencer, and Ochoa 1983; (9) Wall and co-workers.

tribute to the central excitatory state and influence the effect of peripheral stimuli to what, at present, is usually termed the reflex automatism of the isolated spinal cord.

The first reflexes to return in man are the anal and bulbo-cavernous reflexes. Guttman (1976) has drawn attention to the variability in reflex return following spinal shock. One explanation may lie in the residual degree to which non-obvious long descending fibres contribute to the central excitatory state of the apparently isolated spinal cord, as demonstrated by Dimitrijevic *et al.* (1984).

Another important feature of recovery is not only the presence of abnormal reflexes and the time sequence of their return, but also the demonstration of an altered receptive field so that reflexes may be elicited by minimal stimulation of the skin from a progressively wider area (Wall 1980). This enlargement of the receptive field is perhaps the most obvious evidence of alteration of the intact nervous system in response to a lesion, and is probably due to the unmasking of synapses.

As the first stage of spinal shock subsides, afferent impulses from the periphery (skin, tendons, muscles and ligaments, joints, viscera) begin to elicit an excitatory influence. It is as though the process of spinal shock, or, more precisely, the section of sufficient descending pathways, has imposed a resistance on lower spinal cord synapses and in some, at present inexplicable, way this resistance is now lifted (see Illis 1963). Reflexes return in a sequential way and not simultaneously. During the early return of reflexes the stimulus must be strong or summated and the response shows easy fatigue.

The reflexes return in a variable way, but are always altered from how they are in the normal condition. Despite the variability in reflex pattern and the variability in timing of return, there are certain dominant patterns (Guttman 1976):

(a) *Preponderance of the extensor* hallucis longus *and the* tibialis *anterior*. Contraction of the extensor *hallucis longus* on peripheral stimulation is one of the earliest signs of return of reflex activity and constitutes one of the components of Babinski's sign. The *tibialis* anterior reflex contraction belongs to the same group of reflex responses.

(b) *Preponderance of plantar flexors (feet and toes)*.

(c) *Preponderance of crossed reflexes*. The commonest crossed reflex, following plantar stimulation, is an extensor thrust of the contralateral leg. Alternating withdrawal contractions and extensor thrust may occur spontaneously or with very slight external stimulation. Other crossed reflexes include extensor contractions with adductor contractions, and contraction of paralysed extensors of the fingers (in tetraplegics below C6/7) with plantar stimulation.

(d) *Alternating rhythmic contractions*. These were described by Riddoch (1917) and Foerster (1936) and consist of rhythmic alternating flexion and extension contractions of the paralysed lower limbs. The reflex may be confined to single muscles or may involve muscle groups. Rhythmic contractions of this nature may be eliminated by afferent block and this again emphasizes the part played by afferent activity upon the isolated cord.

(e) *Paradoxical reflexes*. These are reflexes which are produced by the spread of the stimulus to antagonists and occur particularly in cervical and *conus* lesions where there is root or segmental damage at the level of the root lesion.

All these abnormal reflex patterns have one thing in common: the stimulus elicits not the usual response, i.e. the response expected in the intact organism, but an abnormal pattern not seen in the undamaged spinal cord. The abnormal pattern of reflexes, therefore, is a reflection of previously existing pathways whose function is normally masked; i.e. these pathways, in the intact nervous system, are structurally intact but play little or no functional part until degeneration of more significant pathways has occurred. Does the unmasking occur simply through the removal of descending control (whatever that precisely means), through chemical hypersensitivity of undamaged endings, or through sprouting of new terminals? This remains a matter of argument and of interpretation. In addition, the abnormal reflex patterns may represent a central alteration of receptive fields secondary to an alteration in afferent input—which itself may be a reflection of partial paralysis. The clinical effect is that when a person with spinal cord damage is examined, one sees a pattern of neurological deficit and abnormal reflex activity which is a reflection not simply of the *fact* of a lesion being present but also of the CNS adapting to the lesion (see Fig. 4.1). Reflex return may fail completely and is then a result of the vertical lesion involving neurons in the spinal grey matter.

Post-mortem study of patients with spinal cord injury who were thought to have clinically complete lesions during life has shown continuity of nerve fibres (Kakulas and Bedbrook 1976). This has been investigated physiologically by Dimitrijevic *et al.* (1984). They found that residual structures in the injured part of the cord can transmit to segments below the lesion, and they also demonstrated supraspinal activation of motor units in paralysed legs during reinforcement manœuvres. By the use of such techniques it should be possible to categorize patients much more accurately. From the clinical point of view it is now important, rather than describing a patient as complete or incomplete at a particular level of the neuraxis, to try to categorize the dysfunction and to discover whether or not the patient can modify or augment or suppress the dysfunction; i.e. to find out whether the segmental function below the lesion can or cannot be influenced by the brain. Residual long descending fibres may have significance at the present time in terms of the application of stimulation techniques to relieve spasticity (Dimitrijevic *et al.* 1986a, 1986b), and in the future through, perhaps, the exploitation of techniques of nerve grafting (Richardson, McGuiness, and Aguayo 1982).

The inference from biological models also lends some support to the above concept. Broadly speaking, there are two types of model of nervous integration—those with a random or diffuse structure, and those with a well-localized or specific structure. The rapid association of sensory information coming from widely differing sources and of differing modalities requires a richly cross-connected nervous system. However, the existence of highly specific inborn circuits in the CNS implies the genetic determination of synaptic connections and reduces to a minor role the effect of use and disuse on connectivity. Complete randomness of circuitry is contrary to known anatomical facts, but, on the other hand, genetically rigid circuitry could not possibly subserve the multitude of genetically unpredictable activities of an organism during its lifetime. In the random type of model, non-specificity of organization will itself provide stability against disturbance. There is, however, a sharp transition from stability to instability as the connectance reaches a critical level (Gardner and Ashby 1970; May 1972). In the CNS this potential instability would be guarded against because any specific circuits would reduce the randomness of the structure. If only specific circuits are present then stability does not exist in the presence of any outside perturbation. The simplest form of a diffusely structured nervous system is where the probability of any one unit of the system being connected to any other unit is equal. Assuming the property of synaptic facilitation, such a structure could be altered as the result of a sensory input and perform a variety of learned tasks (Harth and Edgar 1967). Some specificity could be established as a result of repetitive stimulation, since the effect of such stimulation is to produce a demonstrable structural change in appropriate synapses (Bazanova, Evdokimov, Maiorov, Merkulova, and Chernigovskii 1966; Illis 1969).

Blakemore and van Sluyters (1975) investigated the development of the kitten's visual cortex and suggested that the general synaptic field was laid down on the basis of genetic information. Upon this genetically determined field, visual experience produced alterations: 'the experience itself provides essential information that cannot be accurately forecast by genetic messages alone'.

The acceptance of this type of random–specific model as a means of studying the functioning of the CNS assumes the relative unimportance of connections, and indeed of the nerve cells themselves, as regards integration at least, and throws into prominence the areas of connectivity or areas of discontinuity of the nervous system. What becomes important in such a system is not the fact of the connection between units, but rather the type of connection made. This concept of relative contribution made by pathways is implicit in some of Sherrington's writings on spinal shock—spinal shock is not due to 'the mere wound but to the number . . . of descending nerve paths through which the lesion breaks' (Sherrington 1910), and in much of Pavlovian physiology, which suggests that recovery of function may be due to an opening-up of paths not normally or previously used. The relative contribution of a pathway to a synaptic zone is not rigid but can be altered by use, by disuse, by degeneration and regeneration, and by the effect of toxins (see Illis 1973c). The clinical interpretation of such a model is that any perturbation of such a system—for example, by destroying one set of inputs to the network—could result, not in the breakdown of the whole network but in the reorganization of the network. The result in terms of reflexes has been discussed above. The abnormal pattern of reflexes, or pathological reflex spread, changes with time, and its variation depends on the maturity of the organism's CNS at the time the lesion was sustained. The pattern of abnormal reflexes reflects anatomical and physiological changes in the spinal cord and pharmacological changes which are a result of the lesion. More detailed study should enable us to influence this process and to influence recovery.

References

Abbate, A. D., Cook, A. W., and Attalah, N. (1977). The effect of electrical stimulation of the thoracic spinal cord on function of bladder in multiple sclerosis. *J. Urol.* **117**, 285–8.

Aitken, J. T. and Bridge, J. E. (1961). Neuron size and neuron population density in the lumbosacral region of the cat's spinal cord. *J. Anat.* **95**, 38–53.

Armstrong, J. and Young, J. Z. (1957). End feet in the cerebral cortex. *J. Physiol. (Lond.)* **137**, 10–11.

Armstrong, J., Richardson, K. C., and Young, J. Z. (1956). Staining end feet and mitochondria after postchroming and carbowax embedding. *Stain Technol.* **31**, 263–70.

Bantli, H., Bloedel, J. R., and Thienprasit, P. (1975). Supraspinal interactions resulting from experimental dorsal column stimulation. *J. Neurosurg.* **42**, 296–300.

Barolat-Romana, G., Muklebust, J. B., Hemmy, D. C., Myklebust, R., and Wenninger, W. (1985). Immediate effects of spinal cord stimulation in spinal spasticity. *J. Neurosurg.* **62**, 558–62.

Baskin, D. S. and Hosobuchi, Y. (1981). Naloxone reversal of ischaemic neurological deficits in man. *Lancet* **ii**, 272–5.

Baskin, D. S., Keick, C. F., and Hosobuchi, Y. (1982). Naloxone reversal of ischaemic deficits in baboons is not mediated by systemic effects. *Life Sci.* **31**, 2201–4.

Bazanova, I. S., Evdokimov, S. A., Maiorov, V. N., Merkulova, O. S., and Chernigovskii, V. N. (1966). Morphological and electrical changes in interneuronal synapses during passage of rhythmic impulses. *Fed. Proc.* **25**, T187–90.

Bennett, E. L., Rosenzweig, M. R., and Wu, S. Y. C. (1973). Excitant and depressant drugs modulate effects of environment on brainweight and cholinesterase. *Psychopharmacologia (Berlin)* **33**, 309–28.

Berg, V., Bergmann, S., Hovdal, H., Hunstead, N., Johnson, H. J., Levin, L., and Sjaastad, O. (1982). The value of dorsal column stimulation in multiple sclerosis. *Scand. J. Rehab. Med.* **14**, 183–91.

Bernstein, J. J. and Wells, M. R. (1980). Puromycin induction of transient regeneration in mammalian spinal cord. *Prog. Brain Res.* **53**, 21–38.

Blakemore, C. and van Sluyters, R. C. (1975). Innate and environmental factors in the development of the kitten's visual cortex. *J. Physiol.* **248**, 663–716.

Brandt, N. J., Terenius, L., Jacobsen, B. B., Klinken, L., Nordus, A., Brandt, S., Blegvad, K., and Yssing, M. (1980). Hyper-endorphin syndrome in a child with necrotising encephalopathy. *New Eng. J. Med.* **303**, 914–16.

Brailowski, S. (1980). Neuropharmacological aspects of brain plasticity. In *Recovery of function: theoretical considerations for brain injury rehabilitation* (ed. P. Bach-y-Rita), pp. 187–224. University Park Press, Baltimore, Maryland.

Braun, J. J., Meyer, P. M., and Meyer, D. R. (1966). Sparing of brightness habit in rats following decortication. *J. Comp. Physiol. Psychol.* **61**, 79–82.

Cole, D. D., Sullins, W. R., and Isaac, W. (1967). Pharmacological modifications of the effects of spaced occipital ablations. *Psychopharmacologia (Berlin)* **11**, 311–16.

Cook, A. W. and Weinstein, S. P. (1973). Chronic dorsal column stimulation in multiple sclerosis. Preliminary report. *N. Y. State J. Med.* **73**, 2868–72.

Cotman, C. W., Nieto-Sampedro, M., and Harris, E. W. (1981). Synapse replacement in the adult nervous system of vertebrates. *Physiol. Rev.* **61**, 684–784.

Creed, R. S., Denny-Brown, D., Eccles, J. C., Liddell, E. G. T., and Sherrington, C. S. (1932). *Reflex activity of the spinal cord.* Oxford University Press, London.

Denny-Brown, D. (1951). The frontal lobes and their function. In *Modern trends in neurology, series 1* (ed. A. Feiling), pp. 13–89. Butterworth, London.

Denny-Brown, D. (1960). Motor mechanisms. In *Handbook of physiology, section 1. Neurophysiology, part II* (ed. J. Field), pp. 781–96. American Physiological Society, Washington, DC.

Devor, M. (1982). Plasticity in the adult nervous system. In *Rehabilitation of the neurological patient* (ed. L. S. Illis, E. M. Sedgwick, and H. J. Granville), pp. 44–84. Blackwell Scientific Publications, Oxford.

Devor, M. and Wall, P. D. (1976). Dorsal horn cells with proximal cutaneous receptive fields. *Brain Res.* **118**, 325–8.

Dimitrijevic, M. R. and Sherwood, A. M. (1980). Spasticity: medical and surgical treatment. *Neurology* **30**, 19–27.

Dimitrijevic, M. R., Dimitrijevic, M. M., Faganel, J., and Sherwood, A. M. (1984). Suprasegmentally induced motor unit activity in paralysed muscles of patients with established spinal cord injury. *Ann. Neurol.* **16**, 216–21.

Dimitrijevic, M. R., Illis, L. S., Nakajima, K., Sharkey, P. C., and Sherwood, A. M. (1986a). Spinal cord stimulation for the control of spasticity in patients with chronic spinal cord injury. II. Neurophysiological observations. *Central Nerv. Syst. Trauma* **3**, 145–52.

Dimitrijevic, M. M., Dimitrijevic, M. R., Illis, L. S., Nakajima, K., Sharkey, P. C., and Sherwood, A. M. (1986b). Spinal cord stimulation for the control of spasticity in patients with chronic spinal cord injury. I. Clinical observations. *Central Nerv. Syst. Trauma* **3**, 129–44.

Edds, M. V. (1953). Collateral nerve regeneration. *Quart. Rev. Biol.* **28**, 260–76.

Faden, I., Jacobs, T. P., Mougey, E., and Holaday, J. W. (1981). Endorphins in experimental spinal injury: therapeutic effect of naloxone. *Ann. Neurol.* **10**, 326–32.

Faganel, J. and Dimitrijevic, M. R. (1982). Study of propriospinal interneurone system in man: cutaneous exteroceptive conditioning of stretch reflexes. *J. Neurol. Sci.* **56**, 155–72.

Feeney, D. M. and Gold, G. N. (1980). Chronic dorsal column stimulation: effect of H reflex and symptoms in a patient with multiple sclerosis. *Neurosurgery* **6**, 564–6.

Feeney, D. N., Gonzales, A., and Law, W. A. (1982). Amphetamine, haloperidol and experience interact to affect rate of recovery after motor cortex lesion. *Science* **217**, 855–7.

Feeney, D. M., Hovda, D. A., and Salo, A. A. (1983). Phenoxybenzamine reinstates all motor and sensory deficits in cats fully recovered from sensorimotor cortex ablations. *Fed. Amer. Soc. Exp. Biol.* **42**, 1157.

Foerster, O. (1936). Spinal cord. In *Handbuch der Neurologie. Vol. 5* (ed. O. Bumke and O. Foerster), pp. 1–403. Springer, Berlin.

Foreman, R. D., Beall, J. E., Applebaum, A. E., Coulter, J. D., and Willis, W. D. (1976). Effects of dorsal column stimulation on primate spinothalamic tract neurons. *J. Neurophysiol.* **39**, 534–46.

Frolich, A. and Sherrington, C. S. (1902). Path of impulses for inhibition under decerebrate rigidity. *J. Physiol. (Lond.)* **28**, 14–19.

Fulton, J. E. and Sherrington, C. S. (1932). State of flexor reflexes in paraplegic dog and monkey respectively. *J. Physiol. (Lond.)* **75**, 17–22.

Fung, S. T. and Barnes, C. D. (1984). Locus coeruleus control of spinal cord activity. In *Brainstem control of spinal cord function* (ed. C. D. Baines), pp. 215–55. Academic Press, London.

Gardner, M. R. and Ashby, W. R. (1970). Connectance of large dynamic (cybernetic) systems: critical values for stability. *Nature* **228**, 784.

Goldberger, M. E. and Murray, M. (1974). Restitution of function and collateral sprouting in the cat spinal cord: the deafferented animal. *J. Comp. Neurol.* **158**, 37–54.

Gottleib, G. L., Myklebust, B. M., Stefoksi, D., Groth, K., Kroin, J., and Penn, R. D. (1985). Evaluation of cervical stimulation for chronic treatment of spasticity. *Neurology* **35**, 699–704.

Guttman, L. (1976). Spinal shock. In *Handbook of clinical neurology. Vol. 26* (ed. P. J. Vinken and G. W. Bruyn), pp. 243–62. North-Holland, Amsterdam.

Harth, E. M. and Edgar, S. L. (1967). Association by synaptic facilitation in highly damped neural nets. *Biophys. J.* **7**, 689–717.

Hawkes, C. H., Wyke, M., Desmond, A., Bultitude, M. I., and Kanegaonkar, G. S. (1980). Stimulation of dorsal column in multiple sclerosis. *Br. Med. J.* **1**, 889–91.

Illis, L. S. (1963). Changes in spinal cord synapses and a possible explanation for spinal shock. *Exp. Neurol.* **8**, 328–35.

Illis, L. S. (1964*a*). Spinal cord synapses in the cat: normal appearances. *Brain* **87**, 543–54.

Illis, L. S. (1964*b*). Spinal cord synapses in the cat: reaction at motorneurone surface. *Brain* **87**, 555–72.

Illis, L. S. (1967*a*). The relative densities of monosynaptic pathways to cells and dendrites in the ventral horn. *J. Neurol. Sci.* **4**, 259–70.

Illis, L. S. (1967*b*). The motorneurone surface and spinal shock. In *Modern trends in neurology, series 4* (ed. D. Williams), pp. 53–8. Butterworth, Sevenoaks, Kent.

Illis, L. S. (1969). Enlargement of spinal cord synapses after repetitive stimulation. *Nature* **223**, 76–7.

Illis, L. S. (1973*a*). An experimental model of regeneration in the CNS: I Synaptic changes. *Brain* **96**, 47–60.

Illis, L. S. (1973*b*). An experimental model of regeneration in the CNS: II Glial changes. *Brain* **96**, 61–8.

Illis, L. S. (1973*c*). Regeneration in the central nervous system. *Lancet* **i**, 1035–7.

Illis, L. S. (1982). Determinants of recovery. *Int. Rehab. Med.* **4**, 166–72.

Illis, L. S., Sedgwick, E. M., and Tallis, R. C. (1980). Spinal cord stimulation in multiple sclerosis: clinical results. *J. Neurol. Neurosurg. Psychiat.* **43**, 1–14.

Illis, L. S., Sedgwick, E. M., Ogyar, A. E., and Awadalla, M. A. S. (1976). Dorsal column stimulation in the rehabilitation of patients with multiple sclerosis. *Lancet* **i**, 1383–6.

Illis, L. S., Read, D. J., Sedgwick, E. M., and Tallis, R. C. (1983). Spinal cord stimulation in the United Kingdom. *J. Neurol. Neurosurg. Psychiat.* **46**, 299–304.

Iversen, L. I. (1982). Neurotransmitters and CNS disease. *Lancet* **ii**, 914–18.

Jacobs, J. M. and Love, S. (1985). Qualitative and quantitative morphology of human sural nerve at different ages. *Brain* **108**, 897–924.

Kakulas, B. A. and Bedbrook, G. M. (1976). Pathology of injuries of the vertebral spinal cord—with emphasis on the macroscopic aspects. In *Handbook of clinical neurology. Vol. 25* (ed. P. J. Vinken and G. W. Bruyn), pp. 27–42. North-Holland, Amsterdam.

Katz, B. and Miledi, R. (1962). The nature of spontaneous synaptic potentials in motorneurons of the frog. *J. Physiol. (Lond.)* **162**, 51–2.

Krieckhaus, E. E. (1965). Decrements in avoidance behaviour following mammillothalamic tractotomy in rats and subsequent recovery with d-amphetamine. *J. Comp. Physiol. Psychol.* **60**, 31–5.

Laitinen, L. V. and Fugl-Meyer, A. R. (1981). Assessment of functional effect of epidural electrostimulation and selective posterior rhizotomy in spasticity. *Appl. Neurophysiol.* **45**, 331–4.

Larson, S. J., Sances, A., Riegel, D. H., Meyer, G. A., Dallman, D. E., and Swiontek, T. J. (1974). Neurophysiological effects of dorsal column stimulation in man and monkey. *J. Neurosurg.* **41**, 217–23.

Lasek, R. J. and Hoffmann, P. N. (1976). The neuronal cytoskeleton, axonal transport and axonal growth. Cold Spring Harbor Conference on Cell Proliferation. *Cell Motil.* **3**, 1021–49.

Liu, G. N. and Chambers, W. W. (1958). Intraspinal sprouting of dorsal root axons. *Arch. Neurol. Psychiat. (Chicago)* **79**, 46–61.

Luria, A. R., Naydin, V. L., Tsvetkova, L. S., and Vinarskaya, E. N. (1969). Restoration of higher cortical function following local brain damage. In *Handbook of clinical neurology. Vol. 3* (ed. P. J. Vinken and G. W. Bruyn), pp. 368–433. North-Holland, Amsterdam.

Lynch, G., Deadwyler, S., and Cotman, C. (1973). Postlesion axonal growth produces permanent functional connections. *Science* **180**, 1364–6.

Marshall, J. F. (1984). Behavioural consequences of neuronal plasticity following injury to nitrostriatal dopaminergic neurons. In *Ageing and recovery of function in the CNS* (ed. S. W. Scheff), pp. 101–28. Plenum Press, New York.

May, R. M. (1972). Will a large complex system be stable? *Nature* **238**, 413–14.

Meglio, M., Cioni, R., d'Amico, E., Ronzoni, G., and Rossi, G. F. (1980). Epidural spinal cord stimulation for the treatment of neurogenic bladder, *Acta Neurochir.* **54**, 191–9.

Melzack, R. and Wall, P. D. (1965). Pain mechanisms: a new theory. *Science* **150**, 971–9.

Merrill, E. G. and Wall, P. D. (1978). Plasticity of connections in the adult nervous system. In *Neuronal plasticity* (ed. C. W. Cotman), pp. 97–111. Raven Press, New York.

Murray, J. G. and Thompson, J. W. (1957). The occurrence and function of collateral sprouting in the sympathetic nervous system of the cat. *J. Physiol.* **135**, 133–62.

Nakamura, S. and Tsubokawa, T. (1985). Evaluation of spinal cord stimulation for post-apopletic spastic hemiplegia. *Neurosurgery* **17**, 253–9.

Purves, D. and Nja, A. (1978). Trophic maintenance of synaptic connections in autonomic ganglia. In *Neuronal plasticity* (ed. C. W. Cotman), pp. 27–47. Raven Press, New York.

Raisman, G. (1969). Neuronal plasticity in the septal nuclei of the adult rat. *Brain Res.* **14**, 25–48.

Read, D. J., James, E. D., and Shaldon, C. (1985). The effect of spinal cord stimulation on idiopathic detrusor instability and incontinence: a case report. *J. Neurol. Neurosurg. Psychiat.* **48**, 832–4.

Read, D. J., Matthews, W. D., and Higson, R. H. (1980). The effect of spinal cord stimulation on patients with multiple sclerosis. *Brain* **103**, 803–33.

Richardson, P. M., McGuiness, U. M., and Aguayo, A. J. (1982). Peripheral nerve autografts to the rat spinal cord: studies with axonal tracing methods. *Brain Res.* **237**, 147–62.

Riddoch, G. (1917). The reflex function of the completely divided spinal cord in man compared with those associated with less severe lesions. *Brain* **40**, 264–402.

Saade, N. E., Tabet, M. S., Atweh, S. F., and Jabbur, S. J. (1984). Modulation of segmental mechanisms by activation of a dorsal column brainstem spinal loop. *Brain Res.* **310**, 180–4.

Scerrati, M., Onofrj, M., and Pola, P. (1982). Effects of spinal cord stimulation on spasticity: H-reflex study. *Appl. Neurophysiol.* **45**, 62–7.

Schaumberg, H. H., Spencer, P., and Ochoa, J. (1983). The ageing human peripheral nervous system. In *The neurology of ageing* (ed. R. Katzman and R. D. Terry), pp. 111–22. F. A. Davis, Philadelphia, Pennsylvania.

Scheff, S. W., Anderson, K., and Dekosky, S. T. (1984). Morphological aspects of brain damage. In *Ageing and recovery of function in the CNS* (ed. S. W. Scheff), pp. 57–85. Plenum Press, New York.

Sedgwick, E. M., Illis, L. S., Tallis, R. C., Thornton, A. R. D., Abraham, P., El-Negamy, E., Docherty, T. B., Soar, J. S., Spencer, S. C., and Taylor, F. M. (1980). Evoked potentials and contingent negative variation during treatment of multiple sclerosis with spinal cord stimulation. *J. Neurol. Neurosurg. Psychiat.* **43**, 15–24.

Sherrington, C. S. (1910). *The integrative action of the nervous system.* Constable, London.

Sherwood, A. M. (1985). Electrical stimulation of the spinal cord in movement disorders. In *Neural stimulation. Vol. I* (ed. J. B. Myklebust, J. F. Cusick, J. Sances, and S. T. Larson), pp. 111–46. CRC Press, Boca Raton, Florida.

Siegfried, J., Krainick, J. U., Haas, H., Adoriani, C., Meyer, M., and Thoden, U. (1978). Electrical spinal cord stimulation for spastic movement disorder. *Appl. Neurophysiol.* **41**, 134–41.

Shimoji, K., Shimizu, H., Maruyama, Y., Matsuki, M., Kurirayashi, H., and Fujioka, H. (1982). Dorsal column stimulation in man: facilitation of primary afferent depolarization. *Anaesth. Analg.* **61**, 410–13.

Tallis, R. C., Illis, L. S., Sedgwick, E. M., Hardwidge, C., and Garfield, J. S. (1983). Spinal cord stimulation in peripheral vascular disease. *J. Neurol. Neurosurg. Psychiat.* **46**, 478–84.

Tallis, R. C., Illis, L. S., Sedgwick, E. M., Hardwidge, C., and Kennedy, K. (1983). The effect of spinal cord stimulation upon peripheral blood flow in patients with chronic neurological disease. *Int. Rehab. Med.* **5**, 4–9.

Teasdell, R. D. and Stavraky, G. W. (1953). Responses of de-afferented spinal neurons to corticospinal impulses. *J. Neurophysiol.* **16**, 367–75.

Thoden, U., Krainick, J.-U., Strassburg, H. M., and Zimmerman, H. (1977). Influence of dorsal column stimulation on spastic movement disorders. *Acta Neurochir.* **39**, 233–40.

Von Monakow, C. (1914). *Das Grosshirn und die Abbaufunktion durch Kortikale.* Bergman, Herde Wiesbaden, FRG.

Wall, P. D. (1980). Mechanisms of plasticity of connections following damage in adult mammalian nervous systems. In *Recovery of function: theoretical considerations for brain injury rehabilitation* (ed. P. Bach-y-Rita), pp. 91–105. University Park Press, Baltimore, Maryland.

Ward, A. A. and Kennard, M. A. (1942). Effect of cholinergic drugs on recovery of function following lesions of the central nervous system in monkeys. *Yale J. Biol. Med.* **15**, 189–228.

Watson, C. W. and Kennard, M. A. (1945). The effect of anti-convulsant drugs on recovery of function following cerebral cortical lesions. *J. Neurophysiol.* **8**, 221–31.

Weisendanger, M., Chapman, C. E., Marini, G., and Schorderet, D. (1985). Experimental studies of dorsal cord stimulation in animal models of spasticity. In *Clinical neurophysiology in spasticity* (ed. P. J. Delwaide and R. R. Young), pp. 205–19. Elsevier, Amsterdam.

Wolfe, A. (1940). A method of shortening the duration of lower motor neuron paralysis by cholinergic facilitation. *J. Nerv. Ment. Dis.* **92**, 614–22.

Wyckoff, R. W. G. and Young, J. Z. (1956). The motorneurone surface. *Proc. R. Soc.* **B139**, 440–50.

Young, W., Flamm, E. S., Demopoulos, H. G., Tomasula, J. H., and Decrescito, V. (1981). Effects of naloxone on post traumatic ischaemia in experimental spinal contusion. *J. Neurosurg.* **55**, 209.

5

Neuroplasticity and injury to the spinal cord
Clifford J. Woolf

Introduction

Injury to the central nervous system (CNS) is invariably associated both with a loss of function and with a degeneration of neural elements. In this respect the CNS differs fundamentally from the peripheral nervous system (PNS) where injury can be followed by regeneration, repair, and restoration of function. While the CNS does have the capacity to compensate for a loss or disturbance of input, a key question for all those concerned with the clinical management of patients suffering from injury to the CNS is whether, over and above such compensation, the CNS has any capacity for genuine regrowth, whether such regrowth can result in the establishment of functionally significant synapses, and whether these changes offer any help to the patient.

This chapter will present a review based on the literature available up to November 1985 on the topic of plasticity in the nervous system, differentiating between developmental plasticity, functional plasticity, and plasticity following injury to the nervous system. It will be shown that, while great progress has been made in investigating neuroplasticity and its mechanisms, considerably more understanding is required before any realistic assessment can be made as to whether the phenomenon of neuroplasticity offers any hope of treating patients with spinal injury.

Neuroplasticity has been defined in many different ways. A recent definition by Bloom (1985) is that neuroplasticity represents 'those adaptive mechanisms by which the nervous system restores itself towards normal levels of functioning after injury'. This definition views plasticity in terms of a restoration of function. However, while such an optimistic approach may be valid when discussing injuries to the PNS, there is no evidence that injury to the CNS results in changes, other than by compensation, that actually restore function. Unfortunately, the contrary clinical evidence prevails, namely that injury to the CNS may result in maladaptive changes producing, for example, the dysthesia and spasticity present in many patients with spinal injury (Tator 1984).

If a study of neuroplasticity in the CNS is to be useful it should encompass a broader perspective than merely a restoration of function, which may rarely, if ever, occur. For the purposes of this chapter I have defined neuroplasticity as the capacity of the nervous system to modify itself. Such a modifiability includes changes both in structure and in function and can therefore operate both at the molecular and at the gross levels.

Three general categories of such a neuroplasticity exist: developmental plasticity, functional plasticity, and plasticity in response to injury.

Developmental plasticity

For the adult nervous system to develop from the neural tube and neural crest cells, profound changes in the primitive cells of the nervous system must occur. It is beyond the scope of this chapter to discuss the processes of division and differentiation that take place in primitive neurons, other than to state that division ceases before the neuron extends its processes. One major component of developing neurons, first recognized by Cajal (1909), is the presence of growth cones at the tips of their processes, because these represent the instruments of growth and connectivity. Considerable progress has been made in understanding the factors responsible for and affecting the outgrowth of processes from neurons by studying neurite growth in cultured embryonic nerve cells (Bray, Thomas, and Shaw 1978). This is a key area of research, because, without a knowledge of the sequence of events that initiates neurite outgrowth, maintains the neurite, determines its course and extension, and, finally, enables functional contacts to be made between appropriate neurons, we are unlikely to understand structural plasticity in either the immature or the adult CNS.

The neurite appears to grow by the addition of microtubules and neuro-filaments, manufactured in the cell body, to its proximal end, producing an extension of about 1 mm per day, while new membrane is added at the tip of the growth cone (Johnston and Wessels 1980). The neurite has three important properties: it adheres to surfaces, it elongates, and it can be guided. The elongation of the neurite appears to depend upon actin filaments, while its maintenance is dependent on microtubules. The factors affecting guidance of growth cones are poorly understood, but both contact guidances, mediated in part by a cell-surface glycoprotein cell-adhesion molecule (mCAM), and trophic factors, such as nerve growth factor, are involved (Rutishauser, Gall, and Edelman 1978; Greene and Shooter 1980). The guidance of neurites along favoured substrate pathways is assisted in the CNS by radial glial bridges (Rakic 1979) and in the PNS by Schwann cells and basement membranes (Sanes 1983). Major unanswered questions concern how neurite extension is triggered and how growth cones reorganize the proper target and form synapses, and, in particular, whether the patterns of connectivity established reflect intrinsic chemical properties of neurons;

in other words, whether a particular growing neuron will only recognize a specific target neuron (Katz, Labek, and Nauta 1980). One factor of considerable importance is that if a growing neuron fails to make connections it dies (Jacobson 1978), possibly because of the absence of a trophic factor from its target neuron.

The processes determining the outgrowth of neurites, their direction, and their connectivity appear to be a combination of genetic factors and epigenetic influences in the local environment. However, the development of the nervous system also depends upon an external environmental input which can result in either synaptogenesis or synapse elimination. This has been most extensively studied in the visual cortex where the visual connections in young mammals have been shown to be modified by visual input. For example, the distribution of normal ocular dominance columns in the striate cortex is dependent on a visual input to both eyes within a critical period (Wiesel 1982). One possibly general rule that has emerged from such studies is that when two synapses on a target neuron in the immature brain fire synchronously a convergent connection is established, but when the synapses fire asynchronously the two synapses compete until one dominates and active elimination of the non-dominant synapse occurs (Purves and Lichtman 1980). Therefore, there is a constant remodelling of the immature nervous system based on an environmental input assisting in establishing the connections found in the adult brain.

The capacity of undifferentiated or growing neurons to develop new connections by the elongation of processes may not exist in mature differentiated nerve cells. Injured neurons may, however, as part of their response to the injury, dedifferentiate, re-establishing a growing mode. Because of this, the understanding of developmental changes in the CNS is not only of great interest in itself but may also hold answers to some of the major questions concerning the regenerative capacity of the nervous system. It is possible, however, that the local environmental factors that permit growth and guidance of nerve cell processes in the developing state never occur in the adult, even after injury. A recent experiment by Schwab and Thoenen (1985) illustrates this. They used the sciatic and optic nerves as bridges between culture chambers in which dissociated new-born rat sympathetic or sensory nerves had been planted and were growing under the influence of nerve growth factor (NGF). No axons were found to grow into the optic nerve, only into the sciatic nerve. Therefore it is not the absence of trophic factors that prevents growth but something about the local microenvironment in the sciatic nerve that is not present in the optic nerve. One of the differences between the optic and sciatic nerves is the presence of basal laminae in the latter. One of the constituents of the basal lamina, laminin (Rogers, Letourneau, Palm, McCarthy, and Furcht 1983), is an excellent substrate for the growth of neurites. In addition to the absence of optimal growth substrates in the CNS there may also be inhibitory substrate molecules present which prevent growth from taking place (Schwab and Thoenen 1985). Great efforts must be made to study the microenvironment of the developing CNS to see how it differs from that in the adult.

Functional plasticity

I have used this term to describe changes in the adult CNS produced by changes in environmental input. Unlike the immature CNS where a major remodelling occurs, the changes in the adult nervous system are more subtle. However, the capacity to store and retrieve information (i.e. memory) and the ability to use such information to modify behaviour (i.e. learning) indicates that the adult nervous system is not fixed and immutable. The adult CNS appears to be able to modify its response properties over several time epochs. The shortest of these is the millisecond range in response to the classical fast transmitters. These transmitters, which are likely to be glutamate, aspartate, and acetylcholine for excitations, and glycine, gamma aminobutyric acid (GABA), and the monoamines for inhibitions, modify membrane ion channels, altering conductances and thereby producing short-lasting voltage disturbances. The synapse is the major site for functional or structural changes and this could operate by modifying the fast transmitter systems. Such changes could occur pre- or post-synaptically. Pre-synaptic changes could include an increase or decrease in the availability of the transmitter, a change in the amount of transmitter release by a given depolarization of the axon terminal (e.g. by a change in the state of phosphorylation of synapsin; Nestler and Greengard 1983), or by a change in the number of pre-synaptic autoreceptors. Post-synaptic changes could include a change in the number of receptors, in the efficacy of the receptors, in the properties of the ion channels, and even in the structure of the post-synaptic site, such as a dendritic spine which may be able to alter the effect of synaptic potentials on the soma. The one area of the brain that has been extensively studied, in terms of the capacity of certain conditioning inputs to produce prolonged alteration in synaptic efficacy, is the hippocampus. Brief high-frequency stimulation of the perforant pathway produces a prolonged facilitation of subsequent inputs along this pathway.

In the spinal cord similar relatively short-lasting alterations in function after repeated stimulation have also been described. These include both habituation and potentiation (Mendell 1984). In a molluscan model of habituation the phenomena have been shown to result from a pre-synaptic calcium current inactivation for habituation and from an action on a potassium channel for sensitization (Klein, Shapiro, and Kandel 1980). No data are available on the mechanisms involved in the spinal cord. The monosynaptic reflex made by Ia-afferents on alpha-motoneurons has been most extensively studied, and brief trains of impulses in the Ia-afferents produce a homosynaptic increase in the amplitude of the reflex lasting tens of seconds which also appears to be pre-synaptic in origin (Hirst, Redman, and Wang 1981).

Moving on from the fast transmitter systems to neuromodulators we enter an area where the nervous system has the capacity to modify its function, not in the millisecond time range but in the seconds-to-hours range. It is now recognized that, in contrast to classical dogma, most neurons have more than one neurochemical which is released at their axon terminals. The typical finding is the colocalization of a fast transmitter with a neuropeptide; e.g. 5-hydroxytryptomine (5-HT) with substance P, or acetylcholine with vasoactive intestinal polypeptide (VIP). The neuropeptides appear to function in a mode quite different to that of the fast transmitters, in that instead of producing a short-latency–short-duration alteration in membrane potential they have the capacity to produce prolonged subthreshold effects which will modify the action of the fast transmitter. One example in the spinal cord of the action of neuromodulators is the sequence of changes that follow peripheral tissue injury. Injury results in animals in an increase in the excitability of the spinal cord, manifesting as a decrease in the threshold and as an increase in the responsiveness of the flexion reflex (Woolf 1983, 1984). Such changes appear to be the result of an input into the spinal cord from C-primary afferents activated by the injury. This C-fibre input produces a prolonged (several hours) heterosynaptic facilitation of the flexor reflex (Wall and Woolf 1984). The mechanism involved appears to be the release from the C-fibres of a neuromodulator which increases the excitability of the appropriate target second-order neurons (Woolf and Wall 1986). These results show that certain categories of sensory inputs can produce alterations in the stimulus–response relations of the spinal cord which outlast the stimulus period. We have indirect evidence that neuropeptides such as substance P and calcitonin-gene-related peptide may be involved (Woolf and Wiesenfeld-Hallin 1986). Why the neuropeptides produce longer-lasting effects than the amino-acid transmitters is not known. One possibility is that the peptides via the post-synaptic receptors alter the levels of second messengers in dorsal horn neurons. These second messengers, such as Ca^{2+}, inositol trisphosphate (IP_3), cyclic AMP, cyclic GMP, and diacylglycerol, could then, via the activation of substrate-specific protein kinases, result in the phosphorylation of specific membrane proteins, altering their function (Nestler and Greengard 1983). Another interesting feature of the neuro-modulators is that they may not act in the spatially restricted point-to-point contact way that classical fast transmitters do, but may diffuse from their site of release and act on a widely dispersed set of structures (Jan and Jan 1982). The capacity of certain afferent inputs to produce substantial alterations in the metabolic activity of the spinal cord has recently been demonstrated (Woolf, Chong, and Rashdi 1985), indicating that neurotransmitters/neuro-modulators act not only on membrane excitability but that neurons can, via a receptor-mediated change in second messengers, alter the internal chemical milieu of their target neurons.

Whether or not normal environmental inputs can produce structural changes in neurons, and therefore introduce an even longer time course to 'functional plasticity', has been a major debate for many years. Recent studies on sympathetic ganglia show that, even in the mature nervous system, alterations in form do take place (Purves and Hadley 1985). The extent and significance of such changes in the CNS is the subject of major research efforts at present, but no clear answer is available yet.

Taken together, the modifiability of fast transmitter systems at synapses, the prolonged actions of neuromodulators, and the possibility that normal inputs can produce structural changes in the CNS show that the nervous system is not rigid, with a predetermined connectivity, but that it is plastic. This functional plasticity may well contribute to the sequence of reactive events that occur in the CNS following injury, where it is damage to the nervous system and not an environmental input that represents the trigger for changes.

Plasticity following injury to the nervous system

To evaluate the modifications in the nervous system that can occur following injury it is necessary to differentiate between injury to the PNS or CNS and between injury to immature or adult nervous systems.

That regeneration of the peripheral branches of primary afferent neurons and of the axons of motoneurons can occur has long been established (see Sunderland 1968). This fact alone must give us encouragement, in that it means that at least some neurons in the adult have the capacity to respond to injury in a way that can lead to the restoration of function. What it is precisely about peripheral axons that permits regrowth is not absolutely certain. Whether it is the absolute presence of trophic factors that are absent in the CNS, a relatively higher amount of a general trophic factor (Muller, Harter, Hanger, and Shooter 1985), the absence of glial scars (Reier, Stensaas, and Guth 1983), a unique anatomical arrangement of Schwann cells and the basal laminae of endoneural tubes providing the required growth substrate (Freed, de Medinaceli, and Wyatt 1985), the growth potential of primary afferent neurons or motoneurons, or subtle combinations of all these factors is not known.

The relative ease with which studies of injury to the PNS can be made has meant that there is a vast literature on the subject. I will not attempt to review this work in detail and refer the reader to the following excellent reviews: Sunderland 1968; Wall and Devor 1978; Mendell 1984; Freed *et al.* 1985; Selzer 1985. Instead I will list the major changes that occur following injury to primary afferents and to motoneurons, emphasizing recent findings.

Injury to primary afferents

Crushing or sectioning of an axon will immediately produce an injury discharge which will be fired ortho- and antidromically (Wall, Waxman, and Basbaum 1974). The ruptured axon will then be sealed within minutes by a calcium-dependent process (Yawo and Kuno 1985). After these acute changes, changes with slower time courses begin to occur: the degeneration of the distal segment, the development of sprouts, a decrease in conduction velocity, and those series of morphological and chemical changes that constitute the axon reaction of chromatolysis (Lieberman 1971). Whether it is the transfer of some chemical factor from the site of the injury to the cell body that initiates these changes, or the absence of a trophic factor which is normally present, is not yet resolved. However, what is clear is that in addition to the regeneration that may occur following sectioning of a sensory axon, other less-adaptive changes may occur. For example, there is: (1) the development of abnormal membrane excitability in the sprouts, which begin to act as a source for ectopic discharges (Wall and Devor 1978); (2) the decrease in synaptic efficacy of the central terminals of the afferents (Goldring, Kuno, Nunez, and Snider 1980); and (3) cell death (Aldskogius, Arvidsson, and Grant 1985; Tessler, Himes, Artymyshyn, Murray, and Goldberger 1985). The latter appears to occur more in young animals than in older and to a greater extent the closer the injury is to the cell body.

The changes that occur following sectioning of the peripheral branch of a primary afferent neuron are not, however, restricted to that neuron alone. There is evidence that peripheral nerve sectioning produces profound changes in second-order dorsal horn neurons, both in terms of their morphology (Gobel 1984) and their physiology. Peripheral nerve sectioning, for example, produces a reorganization of the somatotopic input to dorsal horn neurons (Devor and Wall 1981a, b; Lisney 1983; Markus, Pomeranz, and Krushelnycky 1984) and to dorsal column nuclei (Kalaska and Pomeranz 1982). These results are of great importance because they indicate that damage to a neuron can affect the function of a second neuron innervated by the first, even when the synaptic contact between the two neurons is maintained; in other words it is not only degeneration that produces changes in the nervous system following injury. The consequences of injury are subtle and spread transsynaptically from the injured cell to uninjured cells. Whether these changes are adaptive, attempting to over-come the effects of deafferentation, is difficult to establish. There is a contrary argument that some of the clinical syndromes following peripheral nerve injury such as causalsia are the product of central transsynaptic changes in the spinal cord (Wall 1985). Nerve cells do not exist independently but are part of a complex interconnecting network. The communication between neurons does not only involve synaptically mediated modifications of electrical activity, but also involves a more subtle trophic dependence between neurons. A recent example of this is the finding in sympathetic ganglia that the particular type of afferent fibre innervating a post-synaptic cell will determine the receptor type expressed on the membranes of that neuron (Marshall 1985). This means that the genetic expression of a given neuron can be modified by alterations in the neurons that innervate it.

When the central branch of a primary afferent is damaged the central terminals of the afferent degenerate, although the cell body seems unaffected. There is no evidence that the distal ends of the central branches regenerate, although collateral sprouting of uninjured axons has been reported to occur (Liu and Chambers 1958; Illis, 1967, 1973; Goldberger and Murray 1982; Tessler *et al.* 1985). Not all investigators have been able to demonstrate such sprouting though (Kerr 1975; Rustioni and Molennar 1975; Rodin, Sampogna and Kruger 1983), and there is no evidence for such sprouting after peripheral nerve sectioning (Seltzer and Devor 1984). Interestingly, although peripheral nerve sectioning is reported to produce transsynaptic changes in the structure of dorsal horn neurons, no similar changes have yet been observed after dorsal root sectioning (Brown, Busch, and Whittington 1979).

Dorsal rhizotomy, like peripheral nerve sectioning, appears to be associated with an alteration in the somatotopic map of second-order dorsal horn neurons (Basbaum and Wall 1976; Sedevic, Ovelmen-Levitt, Karp, and Mendell 1983), although this is not found by all laboratories (Brown, Brown, Fyffe, and Pubols 1983). Whether the changes are the result of sprouting, the unmaking of previously silent synapses, or the result of increases in the excitability of deafferented cells is still not clarified. Whatever the mechanism, though, these results indicate once again the capacity of the nervous system to reorganize itself after suffering a disturbance to one of its input components. As in the case of peripheral sectioning, the deafferentation resulting from dorsal root sectioning is associated with unpleasant *sequelae* in some patients; e.g. the pain associated with brachial plexus avulsion injury.

The effects of both peripheral nerve sectioning and dorsal rhizotomy are not restricted to changes in the dorsal horn. Changes following dorsal root sectioning have been reported in the thalamus (Wall and Egger 1971) and the cortex (Franck 1980). Following peripheral nerve sectioning substantial changes in the somatotopic organization of the primate sensory cortex have been described by Mezernich, Kaas, Wall, Sur, Nelson, and Fellerman (1983). However, we are still unable to answer the question as to whether these responses to primary afferent input are compensatory functional changes or are due to structural regeneration. On balance the evidence still points to the former explanation. One reason for this is that our views of the nervous system as a functionally hard-wired system are beginning to change. The properties of neurons must now rather be seen as undergoing constant, dynamic reorganization in the face of changing inputs. Whether such dynamic capabilities of neurons are exaggerated or accentuated by injury is an important question. The capacity of neurons to change their properties does not necessarily mean, though, that the overall performance of the system following such changes improves. This may be particularly true after injury where these changes may be inappropriate and dysfunctional.

Injury to the axons of motoneurons

Following sectioning of their axons motoneurons can produce sprouts which ultimate reinnervate muscles, resulting in a restoration of function. Whether or not the motoneuron is connected with its original muscle appears to depend, not upon any intrinsic recognition factors, but on whether the regenerating axon grows along the correct muscle nerve reaching denervated muscle. It is certainly possible for extensor motoneurons to innervate flexor muscles or for motoneurons formerly innervating slow-twitch muscles (type S) to reinnervate the pale fast-twitch muscles. One outcome of such cross-innervation has been the recognition that the properties of muscle fibres (speed, duration, and force of contraction, and their metabolism) is determined by the properties of the motoneuron innervating the muscle. In primates, cross-reinnervation experiments, in which the nerves to flexors and extensors are deliberately swapped, result in the re-establishment of excellent control of motor performance after regeneration has occurred (Brinkman, Porter, and Norman 1983). This appears to be due, not to a change in the function of the flexor and extensor motoneurons which now innervate their antagonists, but to a suppression of their function. The CNS can compensate, therefore, for wrong peripheral connections, and this plasticity may be as important as the regeneration for the restoration of motor function.

Apart from the regrowth of the severed axons, axotomy also produces other changes in motoneurons (see Czeh, Kuno, and Kuno 1977; Mendell 1984). These include an increase in the excitability of the motoneuron membrane due to a decreased duration of the after-hyperpolarization (Gustafssen 1979) and the appearance of dendritic spikes (Heyer and Llinas 1977). These changes in the membrane properties of axotomized motoneurons may be either a part of the alteration of the cell into a regenerative phase or of the dedifferentiation of the cell into one with immature membrane properties. Restoration of contact with muscle results in a return of the membrane to its former state.

In addition to the increase in the excitability of the motoneuron there is also a decrease in monosynaptic transmission (Kuno and Llinas 1970), which appears to be due to a loss of connectivity, particularly of axosomatic Ia-synapses (Mendell, Munson, and Scott 1976). This loss of synaptic input onto an injured neuron once more shows that uninjured neurons can be affected by injury to their target neurons. Apart from these electrophysiological changes the axotomized motoneurons undergo an enormous metabolic change, which may reflect both the response to injury and the passage of the neuron to a growing phase (Woolf *et al.* 1984). A feature of the changes in the axotomized motoneuron is the extent to which these are reversible, even after prolonged (six-month) delays in regeneration (Goldring *et al.* 1980).

Injury to the immature PNS

Two things distinguish injury to the adult PNS and injury to the developing PNS. The first is the greater susceptibility of the axotomized developing neurons to cell death and the second is the greater reorganization such injuries provoke in the CNS (Kalaska and Pomeranz 1979; Waite and Cragg, 1982; Kaas, Merzenich, and Killackey 1983). Therefore, while injured nervous tissue is less likely to survive in the developing nervous system, adjacent uninjured tissue can compensate to a greater extent than in the adult including the introduction of structural changes in the system. The model that has been used most extensively to study such phenomena is the innervation of the whiskers of the rat and mouse by parts of the trigeminal system, and it is changes in the clearly defined whisker barrels in the cortex that are looked for (Woolsey 1984). Unlike the adult, injury to a peripheral nerve in a neonatal animal does result in structural changes in the dorsal horn due to sprouting of uninjured neurons into the area vacated by the injured cells (Fitzgerald 1985). This finding once more reinforces the idea that there are fundamental differences between the immature and adult nervous systems.

Injury to the CNS

Cajal, the great Spanish neuroanatomist, had clear views on the possibility of regeneration in the adult CNS: 'nerve paths are fixed, ended, immutable. Everything may die, nothing may regenerate' (Cajal 1928). While he recognized that regeneration could occur in the adult CNS, this was, he maintained, 'abortive regeneration'. In spite of the optimistic pronouncements of many contemporary neuroscientists, to a large extent, and certainly from the clinical point of view, Cajal's views remain valid today. The CNS is essentially incapable of responding to injury in a regenerative capacity that can convincingly be shown to restore function. What has changed since Cajal's time is both our perspective about why this is so and our prospects of changing this situation. Under certain experimental conditions significant plasticity of the adult CNS can now be definitely demonstrated. While this is most encouraging, we must resist all temptations to predict where this may lead in the future, in terms of clinical management, when such predictions are based more on science fiction than on analysis of the data available. What we must do is try to understand how the nervous system responds to injury and how we can alter this response in an attempt to promote functional recovery. New developments offer us exciting possibilities, but they can only evolve from a thorough understanding of the mechanisms brought into play.

The effect of injury to the spinal cord has been extensively reviewed (see Windle 1980; Kao, Bunge and Reier 1983). It still remains true that myelinated fibres are unable to regenerate through the site of a lesion and that this is not due simply to the mechanical obstruction generated by the scar tissue resulting from glial and connective tissue elements. While such glial scars may be important (Reier *et al.* 1983), we must also take into account the intrinsic capabilities of differentiated neurons to form growth cones, their ability to elongate for sufficient distances to bridge a lesion, and the capability of an injured neuron to re-establish contact with its former target neurons.

When unmyelinated axons are damaged chemically and not mechanically, sprouting of the axons does occur (Nobin, Baumgarten, Björklund, Lachenmayer, and Stenevi 1973). In the adult, lesions to the corticospinal tract do not result in collateral sprouting (Kucera and Wiesendanger 1985), and the recovery of motor function must be the result of a reorganization of existing circuitry. In the neonate, lesions distal to a developing pathway are overcome by the descending fibres bypassing the lesions to form their appropriate connection (Bernstein and Stelzner 1983; Shreyer and Jones 1983). In spite of this there is no evidence that injury to the spinal cord results in axonal growth across the transection (Gearhart, Oster-Gracite, and Guth 1979; Cummings, Bernstein, and Stelzner 1981). Spinal transection, in addition to severing contact between the brain and spinal cord, can result in substantial cell death in the axotomized descending pathways (Goshgarian, Koistinin, and Schmidt 1983). There are also differences in the capabilities of the axons at different neurons to develop growth cones after they have been severed (Fishman and Kelly 1984).

Below the site of a half or complete spinal cord section substantial anatomical and physiological reorganization occurs. Such lesions in neonates result in sprouting of primary afferents (Hulsebosch and Coggeshall 1983), and in neonates the recovered ability to use their hindlimbs is substantially greater than in the adult (Stelzner, Ershler, and Weber 1975;

Stelzner, Weber, and Prendergast 1979). In the adult the capacity for sprouting is diminished (Bregman and Goldberger 1983) but not absent (Murray and Goldberger 1974). Functionally major changes occur in the reflex organization of the spinal cord, with changes in visceral reflexes (de Groat, Booth, Milne, and Roppolo 1982), respiratory motoneurons (Kirkwood, Sears, and Westgaard 1980), the flexor reflex (Fijimori, Kato, Matushima, Mori, and Shimamura 1966), and Ia-monosynaptic reflexes (Nelson and Mendell 1979); whether these changes are due to sprouting into vacated synaptic sites or the results of changes in the membrane properties of motoneurons and interneurons is not certain. Denervation supersensitivity, the removal of tonic descending influences, and a disruption in the trophic influences of descending neurons on spinal neurons may all play a role.

Two recent advances have made a fundamental contribution both to the understanding of neuroplasticity and to the possibility of therapeutic interventions. The first is the work of Aguayo and his colleague, which shows that central neurons can regenerate along peripheral nerve grafts, and the second is that of Björklund and his colleagues, which demonstrates that intracerebral neural implants can survive and possibly contribute to the reconstruction of damaged circuitry.

The work of Aguayo leads on from Cajal's hypothesis that the regenerative capacity of neurons is closely dependent on influences arising from the environment surrounding the nerve fibres and that the environment of the PNS differs from that of the CNS (Cajal 1928). To study this, sciatic nerve segments were grafted into the CNS (Richardson, McGuinness, and Aguayo 1982). Using anterograde and retrograde tracing techniques it was shown that intrinsic central neurons have the capacity to grow axons along peripheral nerve grafts (Aguayo 1985). When the *medulla oblongata* and spinal cord of rats were bridged by such grafts at the point of complete transection no functional improvement was found, nor was any extension of the corticospinal pathway into the graft found, although other neurons grow axons into the graft (Richardson *et al.* 1982). Recent work indicates that it is only injured neurons which regenerate into these grafts and not collaterals from uninjured neurons (Friedman and Aguayo 1985). The axons present within the graft are electrically active, but only a few have clear peripheral receptive fields (Munz, Rasminsky, Aguayo, Vidal-Sanz, and Devor 1985). While these findings are very exciting, in that they show that given the right environment neurons in the CNS can grow for prolonged distances, none of the grafts result in a reinnervation of the CNS. The growing axons are simply unable to penetrate the CNS and form functional connections (Aguayo 1985). A major effort must be made into understanding what it is about the CNS microenvironment that prevents further growth and what factors are necessary in order for functional synapses to be established.

We have no data on what are the minimum number of contacts required to produce a substantial functional improvement, nor do we know whether the nervous system will be able to cope with functionally inappropriate connections (e.g. vestibular nuclei to flexor reflex dorsal horn interneurons). If descending axons only make synapses on target neurons that they recognize by surface markers, then the possibility that this approach can be used for clinical treatment is that much greater.

Much more work is required in this important field. We at last have the capacity to provoke injured neurons into a regenerative phase producing growing axons. We must now free them to make functional connections.

Intracerebral grafts offer two possibilities for the treatment of disease or injury to the nervous system (Björklund and Stenevi 1984). The first is that the graft will provide a missing neurochemical which is required for the normal functioning of a system. For this to occur no specific functional contacts need to be established; the grafted cells just pour out sufficient quantities of the desired chemical in close proximity to the target neurons. This approach would be useful for the treatment of conditions like Parkinson's disease, where absence of dopamine from the nigrostriatal pathway is responsible for much of the symptomatology. In the case of spinal injury such an approach may be useful only in providing a neurotransmitter/neuromodulator which could dampen down the excessive excitability of the spinal cord.

The second possibility is that intracerebral grafts may promote the bridging of regenerating axons across lesions to the brain and spinal cord. This may occur either by providing an appropriate substrate for the regeneration of injured axons, much like the sciatic nerve grafts (Björklund, Gage, Dunnett, and Stenevi 1985), or by the embryonic neurons in the graft actually forming synaptic connections with denervated cells. This latter approach has been used with some success in the hippocampus and neostriatum where functional connections are established between the donor cells and the deafferented neurons and this is associated with an improvement in behaviour (Björklund et al. 1985).

The significance of these findings for spinal cord injury are not yet known. A feature of spinal cord function is that it operates under several distinct command centres, responsible for posture, maintenance of equilibrium, organization of movement, and skilled voluntary motor performance. To operate it requires fast accurate feedback from proprioceptive and other sensory afferents. It is difficult to envisage that embryonic neurons, for argument's sake primary motor cortical neurons injected into the site of a lesion, will offer much in the way of improved motor function, even if they establish connections with the appropriate deafferented motor- and interneurons. In other words, it may be an impossible task to attempt to structurally recreate the CNS after injury by the injection of undifferentiated cells, in regions where the functioning of the CNS is as dependent on complex circuitry, as the circuits that control motor function in the spinal cord.

Conclusions

Progress in science is rarely predictable. Major developments frequently occur unexpectedly and from the most unlikely sources. What was considered impossible at one point in time becomes taken for granted within a relatively short time span. Unfortunately, defining a specific goal does not necessarily accelerate progress in a particular field. Science is not about the application of technology, but rather is concerned with the discovery of new facts.

Our understanding of neuroplasticity has increased enormously in the last decade, but we remain tantalizingly removed from any application of our knowledge in the management of conditions such as spinal injury. In practical terms there is no regenerative capacity of the CNS. We now know, however, from experimental work that this is not an absolute fact. Neuroplasticity does occur in the adult CNS after injury. What we must aim to identify are the factors that enable injured axons to form growth cones, the positive and negative factors promoting and inhibiting elongation of axons, and the factors determining whether a growing axon can make a functional contact, not with any neuron but with an appropriate target. Whether such knowledge will be applicable for clinical application remains to be seen, but without such knowledge all that we will definitely know is that nothing can be done to promote regeneration in the CNS.

References

Aguayo, A. J. (1985). Capacity for renewed axonal growth in the mammalian central nervous system. In *Central nervous system plasticity and repair* (ed. A. Bignani, F. E. Bloom, C. L. Breis, and A. Adelaye), pp. 31–40. Raven Press, New York.

Aldskogius, H., Arvidsson, J., and Grant, G. (1985). The reaction of primary sensory neurons to peripheral nerve injury with particular emphasis on transganglionic changes. *Brain Res. Rev.* **10**, 27–46.

Basbaum, A. I. and Wall, P. D. (1976). Chronic changes in the response of cells in adult cat dorsal horn following partial deafferentation; the appearance of responding cells in a previously non-responsive region. *Brain Res.* **116**, 181–204.

Bernstein, D. R. and Stelzner, D. J. (1983). Plasticity of the corticospinal tract following midthoracic spinal injury in postnatal rat. *J. Comp. Neurol.* **221**, 382–400.

Björklund, A. and Stenevi, V. (1984). Intracerebral neural implants: neuronal replacement and reconstruction of damaged circuitries. *Ann. Rev. Neurosci.* **7**, 279–308.

Björklund, A., Gage, F. H., Dunnett, S. B., and Stenevi, U. (1985). Regenerative capacity of central neurons as revealed by intracerebral grafting experiments. In *Central nervous system plasticity and repair* (ed. A. Bignami, F. E. Bloom, C. L. Breis, and A. Adelaye), pp. 56–62. Raven Press, New York.

Bloom, F. E. (1985). In *Central nervous system plasticity and repair* (ed. A. Bignami, F. E. Bloom, C. L. Breis, and A. Adelaye), pp. 3–11. Raven Press, New York.

Bray, D., Thomas, D., and Shaw, G. (1978). Growth cone formation in cultures of sensory neurons. *Proc. Nat. Acad. Sci., U.S.A.* **75**, 5226–9.

Bregman, B. S. and Goldberger, M. A. (1983). Infant lesion effect. II. Sparing and recovery of function after spinal cord damage in newborn and adult cats. *Brain Res.* **285**, 119–36.

Brinkman, C., Porter, R., and Norman, J. (1983). Plasticity of motor behaviour in monkey with crossed forelimb nerves. *Science* **220**, 438–40.

Brown, A. G., Brown, P. B., Fyffe, R. E. W., and Pubols, L. M. (1983). Effects of dorsal root section on spinocervical tract neurones in the cat. *J. Physiol.* **337**, 589–608.

Brown, P. B., Busch, G. R., and Whittington, J. (1979). Anatomical changes in cat dorsal horn cells after transection of a single dorsal root. *Exp. Neurol.* **64**, 453–68.

Cajal, S. R. (1909). *Histologie du système nerveux de l'homme et des vertébrés.* Maloine, Paris.

Cajal, S. R. (1928). *Degeneration and regeneration of the nervous system.* Oxford University Press, London.

Cummings, J. P., Bernstein, D. R., and Stelzner, D. J. (1981). Further evidence that sparing of function after spinal cord transection in the neonatal rat is not due to axonal generation or generation. *Exp. Neurol.* **74**, 615–20.

Czeh, G., Kuno, N., and Kuno, M. (1977). Membrane properties and conduction velocity in sensory neurones following central or peripheral axotomy. *J. Physiol. (Lond.)* **270**, 165–80.

De Groat, W. C., Booth, A. M., Milne, R. J., and Roppolo, J. R. (1982). Parasympathetic preganglionic neurons in the sacral spinal cord. *J. Auton. Nerv. Syst.* **5**, 23–43.

Devor, M. and Wall, P. D. (1981a). Effect of peripheral nerve injury on receptive fields of cells in cat spinal cord. *J. Comp. Neurol.* **199**, 277–91.

Devor, M. and Wall, P. D. (1981b). Plasticity in the spinal cord sensory map following peripheral nerve injury in rats. *J. Neurosci.* **1**, 679–84.

Fishman, P. S. and Kelly, J. P. (1984). Identified central axons differs in their response to spinal cord transection. *Brain Res.* **305**, 152–6.

Fitzgerald, M. (1985). The sprouting of saphenous nerve terminals in the spinal cord following early postnatal sciatic nerve section in the rat. *J. Comp. Neurol.* **240**, 407–13.

Franck, J. L. (1980). Functional reorganization of cat somatic sensory motor cortex (SMI) after selective dorsal root rhizotomies. *Brain Res.* **186**, 458–62.

Freed, W. J., de Medinaceli, M., and Wyatt, R. J. (1985). Promoting functional plasticity in the damaged nervous system. *Science* **227**, 1544–52.

Friedman, B. and Aguayo, A. J. (1985). Injured neurones in the olfactory bulb of the adult rat grow axons along grafts of peripheral nerve. *J. Neurosci.* **5**, 1616–25.

Fujimori, B., Kato, M., Matushima, S., Mori, S., and Shimamura, M. (1966). Studies on the mechanism of spasticity following spinal hemisection in the cat. In *Muscular afferents and motor control* (ed. R. Granit), pp. 397–413. John Wiley, New York.

Gearhart, J., Oster-Gracite, M. L., and Guth, L. (1979). Histological changes after transection of the spinal cord of fetal and neonatal mice. *Exp. Neurol.* **66**, 1–15.

Gobel, S. (1984). Trans-synaptic effects of peripheral nerve injury. *J. Neurosci.* **4**, 2281–90.

Goldberger, M. E. and Murray, M. (1982). Lack of sprouting and its presence after lesions of the cat spinal cord. *Brain Res.* **241**, 227–39.

Goldring, J. M., Kuno, M., Nunez, R., and Snider, W. R. (1980). Reaction of synapses on motoneurones to section and restoration of peripheral sensory connections. *J. Physiol.* **309**, 185–98.

Goshgarian, H. G., Koistinin, J. M., and Schmidt, E. R. (1983). Cell death and changes in the retrograde transport of horseradish peroxidase in rubrospinal neurones following spinal cord hemisection in the adult rat. *J. Comp. Neurol.* **214**, 251–7.

Greene, L. A. and Shooter, E. M. (1980). The nerve growth factor: biochemistry, synthesis and mechanism of action. *Ann. Rev. Neurosci.* **3**, 353–462.

Gustafsson, B. (1979). Changes in motoneurones electrical properties following axotomy. *J. Physiol. (Lond.)* **293**, 197–215.

Heyer, C. B. and Llinas, R. (1977). Control of rhythmic firing in normal and axotomized cat spinal motoneurones. *J. Neurophysiol.* **40**, 480–8.

Hirst, G. D. S., Redman, S. J., and Wong, K. (1981). Post-tetanic potentiation and facilitation of synaptic potentials evoked in cat spinal motoneurones. *J. Physiol. (Lond.)* **321**, 97–110.

Hulsebosch, C. E. and Coggeshall, R. E. (1983). Age related sprouting of dorsal root axons after sensory deprivation. *Brain Res.* **288**, 77–83.

Illis, L. S. (1967). The motor neuron surface and spinal shock. In *Modern trends in neurology* (ed. D. Williams), pp. 53–68. Butterworth, London.

Illis, L. S. (1973). Experimental model of regeneration in the CNS. *Brain* **96**, 47–60.

Jacobson, M. (1978). *Developmental neurobiology* (2nd edition). Plenum Press, New York.

Jan, L. Y. and Jan, Y. N. (1982). Peptidergic transmission in sympathetic ganglia of the frog. *J. Physiol.* **327**, 219–46.

Johnston, R. N. and Wessels, W. I. (1980). Regulation of the elongating nerve fibres. *Corr. Top. Dev. Biol.* **16**, 165–206.

Kaas, J. H., Merzenich, M. M., and Killackey, H. P. (1983). The reorganization of the somatosensory cortex following peripheral nerve damage in adult and developing mammals. *Ann. Rev. Neurosci.* **6**, 325–57.

Kalaska, J. and Pomeranz, B. (1979). Chronic paw degeneration causes an age-dependent appearance of novel responses from forearm in 'paw cortex' of kittens and adult cats. *J. Neurophysiol.* **42**, 618–33.

Kalaska, J. and Pomeranz, B. (1982). Chronic peripheral nerve injuries alter the somatotopic organization of the cuneate nucleus in kittens. *Brain Res.* **236**, 35–47.

Kao, C. C., Bunge, R. P., and Reier, P. J. (eds.) (1983). *Spinal cord reconstruction.* Raven Press, New York.

Katz, M. J., Labek, R. J., and Nauta, H. J. W. (1980). Ontogeny of substrate pathways and the origin of the neural circuit pattern. *Neuroscience* **5**, 821–33.

Kerr, F. W. L. (1975). Neuroplasticity of primary afferents in the neonatal cat and some results of early deafferentation on the trigeminal spinal nucleus. *J. Comp. Neurol.* **163**, 305–28.

Kirkwood, P. A., Sears, T. A., and Westgaard, R. H. (1980). Motor performance following partial central deafferentation of motoneurones (abstr.). *J. Physiol. (Lond.)* **312**, 42–43P.

Klein, M., Shapiro, E., and Kandel, E. R. (1980). Synaptic plasticity and the modulation of the Ca^{2+} current. *J. Exp. Biol.* **89**, 117–57.

Kucera, P. and Wiesendanger, M. (1985). Do ipsilateral corticospinal fibres participate in the functional recovery following unilateral pyramidal lesions in monkeys? *Brain Res.* **348**, 297–303.

Kuno, M. and Llinas, R. (1970). Enhancement of synaptic transmission by dendritic potentials in chromatolysed motoneurones of the cat. *J. Physiol. (Lond.)* **210**, 807–21.

Lieberman, A. R. (1971). The axon reaction: a review of the principal features of perikarial responses to axon injury. *Int. Rev. Neurobiol.* **14**, 49–121.

Lisney, S. J. W. (1983). Changes in the somatotopic organization of the cat lumbar spinal cord following peripheral nerve transection and regeneration. *Brain Res.* **259**, 31–9.

Liu, C. M. and Chambers, W. W. (1958). Interspinal sprouting of dorsal root axons. *Arch. Neurol.* **79**, 46–61.

Markus, H., Pomeranz, B., and Krushelnycky, D. (1984). Spread of saphenous somatotopic projection map in spinal cord and hypersensitivity of the foot after chronic sciatic denervation in adult rat. *Brain Res.* **296**, 27–40.

Marshall, L. M. (1985). Presynaptic control of synaptic channel kinetics in sympathetic neurones. *Nature* **317**, 621–3.

Mendell, L. M. (1984). Modifiability of spinal synapses. *Physiol. Rev.* **64**, 260–324.

Mendell, L. M., Munson, J. B., and Scott, J. G. (1976). Alterations of synapses on axotomized motoneurones. *J. Physiol. (Lond.)* **255**, 67–79.

Mezernich, M. M., Kaas, J. H., Wall, J. T., Sur, M., Nelson, R. J., and Fellerman, D. J. (1983). Progression of change following median nerve section in the cortical representation of the hand in areas 3b and 1 in adult owl and squirrel monkeys. *Neuroscience* **10**, 639–65.

Muller, H. W., Harter, P. J., Hanger, D. H., and Shooter, E. M. (1985). A specific 37,000 Dalton protein that accumulates in regenerating but not in nonregenerating mammalian nerves. *Science* **228**, 499–501.

Munz, M., Rasminsky, M., Aguayo, A. J., Vidal-Sanz, M., and Devor, M. G. (1985). Functional activity of rat brainstem neurons regenerating axons along peripheral nerve grafts. *Brain Res.* **340**, 115–25.

Murray, M. and Goldberger, M. E. (1974). Restitution of function and collateral sprouting in the cat spinal cord. *J. Comp. Neurol.* **158**, 19–36.

Nelson, S. G. and Mendell, L. M. (1979). Enhancement in Ia-motoneuron synaptic transmission caudal to chronic spinal cord transection. *J. Neurophysiol.* **42**, 642–54.

Nestler, E. J. and Greengard, P. (1983). Protein phosphorylation in brain. *Nature* **305**, 583–8.

Nobin, A., Baumgarten, H. G., Björklund, A., Lachenmayer, L., and Stenevi, U. (1973). Axonal degeneration and regeneration of the bulbospinal indolemic neurone after 5,6-dehydroxytryptamine treatment. *Brain Res.* **56**, 1–24.

Purves, D. and Hadley, R. D. (1985). Changes in dendritic branches of adult mammalian neurones revealed by repeated imaging in situ. *Nature* **315**, 404–6.

Purves, D. and Lichtman, J. W. (1980). Elimination of synapses in the developing nervous system. *Science* **210**, 153–7.

Rakic, P. (1979). The role of neuronal glial cell interaction during brain development. In *Neuronal–glial cell interrelationships* (ed. S. A. Sears), pp. 25–78. Springer, New York.

Reier, P. J., Stensaas, L. J., and Guth, L. (1983). The atrocytic scar as an impediment to regeneration in the central nervous system. In *Spinal cord reconstruction* (ed. C. C. Kao, R. P. Burge, and P. J. Reier), pp. 163–98. Raven Press, New York.

Richardson, P. M., McGuinnes, U. N., and Aguayo, A. J. (1982). Peripheral nerve autografts to the rat spinal cord: studies with axonal training methods. *Brain Res.* **237**, 147–62.

Rodin, B. E., Sampogna, S. L., and Kruger, L. (1983). An examination of intraspinal sprouting in dorsal root axons with the tracer horseradish peroxidase. *J. Comp. Neurol.* **215**, 187–98.

Rogers, S. L., Letourneau, P. C., Palm, S. L., McCarthy, J., and Furcht, L. T. (1983). Neurite extension by peripheral and central nervous system neurones in response to substratum around fibionectum and laminin. *Dev. Biol.* **98**, 212–20.

Rustioni, A. and Molennar, I. (1975). Dorsal column nuclei afferents in the lateral funiculus of the cat: distribution pattern and absence of sprouting after chronic deafferentation. *Exp. Brain Res.* **21**, 1–12.

Rutishauser, U., Gall, W. E., and Edelman, G. M. (1978). Adhesion among neural cells of the chick embryo IV. *J. Cell. Biol.* **79**, 382–93.

Sanes, J. R. (1983). Roles of extracellular matrix in neuronal development. *Ann. Rev. Physiol.* **45**, 581–600.

Schreyer, D. J. and Jones, E. G. (1983). Growing and corticospinal axons bypass lesions of neonatal rat spinal cord. *Neuroscience* **9**, 31–40.

Schwab, M. E. and Thoenen, H. (1985). Dissociated neurons regenerate into sciatic but not optic nerve explants in culture irrespective of neurotrophic factors. *J. Neurosci.* **5**, 2415–23.

Sedivic, M. J., Ovelmen-Levitt, J., Karp, R., and Mendell, L. M. (1983). Altered modality convergence after acute and chronic partial deafferentation of spinocervical tract cells in the cat spinal cord. *J. Neurosci.* **3**, 1511–19.

Seltzer, M. E. (1985). Regeneration of peripheral nerve. In *The physiology of peripheral nerve disease* (ed. A. J. Sumner), pp. 358–431. W. .B. Saunders Company, Philadelphia, Pennsylvania.

Seltzer, Z. and Devor, M. (1984). The effect of nerve section on the spinal distribution of neighbouring nerves. *Brain Res.* **306**, 31–7.

Stelzner, D. J., Ershler, W. B., and Weber, E. D. (1975). Effects of spinal transection in neonatal and weanline rats: survival of function. *Exp. Neurol.* **46**, 156–77.

Stelzner, D. J., Weber, E. D., and Prendergast, J. (1979). A comparison of the effect of mid-thoracic spinal hemisection in the neonatal or weanling rat on the distribution and density of dorsal root axons in the lumbosacral spinal cord of the adult. *Brain Res.* **172**, 407–26.

Sunderland, S. (1968). *Nerves and nerve injuries*. Churchill Livingstone, Edinburgh.

Tator, C. H. (1984). Pain following spinal cord injury. In *Neurosurgery* (ed. R. H. Wilkins and S. S. Rengachong), pp. 2368–71. McGraw-Hill, New York.

Tessler, A., Himes, B. T., Artymyshyn, R., Murray, M., and Goldberger, M. H. (1985). Spinal neurons mediate return of substance P immunoreactivity and dorsal root ganglion cell numbers. *Soc. Neurosci. Abstr.* **8**, 305.

Waite, P. M. E. and Cragg, B. G. (1982). The peripheral and central changes resulting from cutting or crushing the afferent nerve supply to the whiskers. *Proc. R. Soc. Lond. B* **214**, 191–211.

Wall, P. D. (1985). Future trends in pain research. *Phil. Trans. R. Soc. Lond. B* **308**, 399–401.

Wall, P. D. and Devor, M. (1978). Physiology of sensation after peripheral nerve injury, regeneration and neurons formation. In *Physiology and pathophysiology of axons* (ed. S. G. Waxman), pp. 377–84. Raven Press, New York.

Wall, P. D. and Egger, M. D. (1971). Formation of new connections in adult rat brain after partial deafferentation. *Nature* **232**, 542–5.

Wall, P. D. and Woolf, C. J. (1984). Muscle but not cutaneous C-afferent input produces prolonged increases in the excitability of the flexion reflex in the rat. *J. Physiol.* **356**, 443–58.

Wall, P. D., Waxman, S., and Basbaum, A. I. (1974). Ongoing activity in peripheral nerve III. Injury discharge. *Exp. Neurol.* **45**, 576–89.

Wiesel, T. N. (1982). Postnatal development of the visual cortex and the influence of environment. *Nature* **299**, 583–91.

Windle, W. F. (1980). *The spinal cord and its reaction to traumatic injury*. Dekker, New York.

Woolf, C. J. (1983). Evidence for a central component of post-injury pain hypersensitivity. *Nature* **306**, 686–8.

Woolf, C. J. (1984). Long term alterations in the excitability of the flexion reflex produced by peripheral tissue injury in the chronic decerebrate rate. *Pain* **18**, 325–43.

Woolf, C. J. and Wall, P. D. (1986). The relative effectiveness of C-primary afferent fibres of different origins in evoking a prolonged facilitation of the flexor reflex in the rat. *J. Neurosci.* **6**, 1433–42.

Woolf, C. J. and Wieselfeld-Hallin, Z. (1986). Substance P and calcitonin gene related peptide synergistically modulate the gain of the nociceptive flexor withdrawal reflex in the rat. *Neurosci. Lett.* **66**, 226–30.

Woolf, C. J., Chong, M. S., and Ainsworth, A. (1984). Axotomy increases glycogen phosphorylase activity in motoneurones. *Neuroscience* **12**, 1261–9.

Woolf, C. J., Chong, M. S., and Rashdi, T. A. (1985). Mapping increased glycogen phosphorylase activity in dorsal root ganglia and in the spinal cord following peripheral stimuli. *J. Comp. Neurol.* **234**, 60–76.

Woolsey, T. A. (1984). The postnatal development and plasticity of the somatosensory system. In *Neuronal growth and plasticity* (ed. M. Kuno), pp. 241–58. Japan Science Society Press, Tokyo.

Yawo, H. and Kuno, M. (1985). Calcium dependence of membrane sailing of the cut end of the cockroach giant axon. *J. Neurosci.* **5**, 1626–32.

6

Neurological and neurophysiological evaluation of spinal cord injury

Wise Young and Peter Mayer

Introduction

Protocols for assessing spinal cord injury depend on the purpose for which data is being collected. If the data is for epidemiological purposes, the classification of spinal cord injury into general categories, such as that proposed by Frankel, Hancock, Hyslop, Melzak, Michaelis, Ungar, Vernon, and Walsh (1967) and others (Cheshire 1970; Lucas and Ducker 1979; Maynard, Reynolds, Fountain, Wilmot, and Hamilton 1979; Chehrazi, Wagner, Collins, and Freeman 1981), will suffice. If clinical therapy is being predicated on the findings, data collection must focus on criteria for the therapy; for example, radiographic analyses of the spinal fracture are needed for planning management of the fracture. If data are being collected to ascertain whether a therapeutic approach alters neurological recovery and to establish criteria for the therapy, studies of a different order of rigour are required.

We will review here approaches developed for documenting treatment effects in spinal cord injury, using the experience gained by the Bellevue Spinal Injury Centre at the New York University (NYU) to exemplify problems encountered in measuring recovery in spinal cord injury. Our aim is not to advocate adoption of a particular protocol or to present the data obtained at NYU, but rather to discuss the requirements for a standard protocol for evaluating treatments of spinal cord injury. A brief historical background will be given first, followed by discussions of the need for detailed and frequent studies, the timing of examinations, the role of neurophysiological testing, the minimum data required, motor scoring, sensory scoring, data analysis methods, and pressing problems that face the development of a standard protocol.

Historical background

Several classification schemes of spinal injured patients have been published. An early system advocated by Guttman (1973) and comprehensively reported by Frankel et al. (1967) utilizes a scale to categorize the

Table 6.1. Three major classification schemes for spinal cord injury; the criteria for each category are paraphrased

Frankel *et al.* (1967)	Maynard *et al.* (1979)	Lucas and Ducker (1979)
Complete sensory–motor loss	Complete lesion	Complete motor
Sensory spared with motor loss	Sensory spared	
Non-useful motor recovery	Useless motor	Partial motor
Useful motor recovery	Useful motor	
Recovery	Ambulatory	Ambulatory

severity of deficits (see Table 6.1). This classification has been widely imitated and applied in subsequent studies; for example, Maynard *et al.* (1979) published a study categorizing spinal injured patients in a similar manner to assess recovery. These approaches broadly generalize recovery criteria and have been useful for documenting the poor prognostic outcome of patients admitted with complete motor and sensory losses. However, they have been less useful for quantifying the degree of recovery, especially in partially recovered patients.

Several groups in the USA have advocated use of scoring systems to document neurological recovery. For example, Ducker and Wallek (1985) and Lucas and Ducker (1979) at the Neurotrauma Center of the Maryland Institute for Emergency Medical Services described an approach in which levels of vertebral trauma were documented radiologically and by motor examination. Ten muscle groups on each side of the body were scored on a scale of 0–5, adding up to 100. Recovery was expressed, in terms of functional change after one year, as a percentage of total possible recovery. Although sensory examinations of the patients were implied, no systematic protocol for scoring sensory changes was described. However, they did introduce four categories of sensory loss: complete, severe partial, partial with caudal sparing, and partial with uniform loss.

Extensive neurological evaluation protocols have been used (Bracken, Webb, and Wagner 1977; Bracken *et al.* 1984, 1985; Bracken and Collins 1985). A multicentre clinical trial funded by the National Institute of Health and administered by Yale University began in 1979. Called the National Acute Spinal Cord Injury Study (NASCIS), this study now involves 10–15 spinal injury centres (Bracken, Shepard, Hellenbrand, Collins, Leo, Freeman, Wagner, Flamm, Eisenberg, Goodman, Perot, Green, Grossman, Meagher, Young, Fischer, Clifton, Hunt, and Rifkinson 1985), including NYU. A major phase-III randomized clinical trial of methylprednisolone was completed in 1984 (Bracken *et al.* 1984). The study protocol stipulated a more detailed neurological documentation of spinal cord injury patients than has been applied in previous studies. The number of muscle groups scored was expanded from the ten recommended by Lucas and Ducker (1979) to fourteen on each side of the body. Sensory responses to light touch and pinprick were recorded for every segmental dermatome. Admission and follow-up examinations at six weeks, six months, and one year after injury were required. Recovery was defined as the difference between admission and follow-up motor and sensory scores.

Need for more detailed and frequent documentation in treatment trials

The trend towards more detailed and frequent assessments of spinal cord injured patients in treatment trials stems from a growing recognition that several different pathological mechanisms can contribute to injury and that many obstacles block recovery (Young 1985). Beneficial effects of therapy directed at one problem may not produce functional return as long as other problems are not resolved. For example, a given treatment may succeed in increasing the number of surviving axons, but having more axons may not yield more function if demyelination of these axons continues to occur. Likewise, it may not be enough to induce and facilitate axonal regeneration. For functional recovery, treatment must also be directed at preparing the potential targets of regenerated axons for synaptic connection, tuning the balance of excitation and inhibition in the target cells so that the axons can produce appropriate responses, and at myelinating the regenerated axons.

No single treatment is likely to achieve all or many of these goals. Multiple treatments aimed at different stages and pathological processes will be necessary. Therefore assessment protocols in clinical trials should possess sufficient sensitivity to detect modest or even transient effects. Certain difficult-to-detect changes, such as improved joint proprioceptivity, descent of the lesion level by a few segments, or more rapid rates of recovery, require careful comprehensive examinations of patients. Assessment protocols that focus only on gross long-term outcomes or on miraculous motor recoveries may miss these important intermediate effects.

Clinical trials also provide limited information. Randomized trials test whether a given treatment is effective, not whether it is optimal. For example, even if a treatment produces significant beneficial effects, the positive results do not necessarily imply that another dose or a different timing and treatment length would not be better. Likewise, if a given treatment turns out to be ineffective, we have not ascertained that another dose or treatment period may turn out to be effective. The evaluation protocol should be designed to detect clues of treatment effects that give insight concerning how to improve the treatment for further studies.

The minimum essential documentation

Theoretically, frequent and comprehensive neurological examination should be done on every spinal injured patient undergoing clinical trials. Practical constraints limit the extent and frequency of data collection on each patient. Developing protocols for multiple institutions imposes another restriction in that resources differ from place to place. Some centres have the luxury of trained neurologists on call to examine patients in great detail daily and instrumentation to monitor patients extensively. Many cannot do so. Thus development of a standard protocol must aim at a minimum set of data that most spinal cord injury centres can acquire.

Since clinical spinal cord injuries vary in injury level and severity, statistically significant results usually cannot be obtained by a direct comparison of treated and control populations. One solution to this problem is to assess changes in function over time, comparing deficits before and after treatment. The pre-treatment examination thus serves as an internal control for each patient. However, since recovery courses in spinal cord injury vary, differing times of pre-treatment examinations may introduce a confounding variable. Thus care must be taken to ascertain the stability of neurological findings prior to treatment. Follow-up examinations should allow enough time for treatment effects to manifest and be frequent enough to catch transient changes.

To document lesion level and severity, some minimum data must be obtained. The lesion level is best determined by a combined motor and sensory examination, defining the level below which deficits are observed. Injury severity is best judged by quantifying the deficits below the lesion level. Thus a description of the spatial distribution of deficits and an accounting of the neurological status below the lesion are required. The examination can be directed at many parameters; e.g. sensory, motor, reflex, bladder, bowel, etc. Since motor and sensory function close to the lesion must be tested to determine lesion level, it is convenient and reasonable to extend the examination to include the areas below the lesion site.

The minimum protocol, therefore, should involve at least two examinations, one before and the other after treatment. The data gathered should reliably provide the lesion level and a quantitative index of injury severity. At NYU, the minimum evaluation protocol that we have instituted for assessing acute spinal cord injury requires documentation of pin and touch sensation in the C4–S5 dermatomes, joint proprioception, motor strength in thirty-two muscle groups, deep tendon reflexes, and bladder function. Patients are examined on the day of admission, six weeks, six months, and one year after injury. In over 50 per cent of the cases, initial examinations were made within twelve hours of injury.

The NASCIS protocol is a subset of the NYU protocol, using the same scoring systems. To this minimum data base, more information can be added. For example, we often evaluate patients more frequently and record cortical somatosensory-evoked potentials on the day of admission, six weeks, six months, and one year after injury. Four sets of four evoked potentials, activated by median nerve and posterior *tibialis* stimulation, are compared with the neurological changes. Other specialized tests can be incorporated; e.g. bladder function, electromyography, etc.

Frequency and timing of assessment

Practical constraints limit the number of times a patient can be studied. At NYU we chose examination times based on the likelihood of detecting functional changes. If a patient is unlikely to change during a period of time, it is wasteful of sparse resources to examine the patient repeatedly over that time. The course of recovery varies from patient to patient, depending on the level and severity of trauma. For our minimum required assessment protocol, we chose time points in which at least 20 per cent of the patients in our population show significant neurological change. For example, our follow-up periods of one week, six weeks, three months, six months, and one year fulfil this criterion. Our population has about a 60:40 ratio of severe:partial injuries. These follow-up times may not be optimal for populations with more severe or partial injuries.

The initial examination is the most important base-line, but also the most variable in terms of timing. For example, 400 patients were admitted to the NYU Spinal Injury Centre within 48 hours after injury. Admission times varied from 2 to 48 hours after injury, although more than 60 per cent were seen within 12 hours and more than 80 per cent within 24 hours. The extent of early recovery depends on the severity of injury. Of the patients whose admission examinations showed no function below the lesion, only 7 per cent recovered function more than three segments below the lesion by a year after injury. In only four of these patients, i.e. 2 per cent, enough sensory and motor function returned to allow independent walking. None recovered within a week. Thus, in severe injuries, the timing of the initial examination within one week after injury is probably not critical.

Since patients admitted with no function below the lesion so seldom recovered function within the first week, it would appear fruitless to examine these patients during this time period. On the other hand, since it is such a rare occurrence, functional return within a week in such patients would represent a significant finding. In contrast, patients with some preservation of function below the lesion may show signs of recovery as early as 48 hours. Treatment may also produce recovery. In one patient treated with a high dose of naloxone (Flamm, Young, Collins, Piepmeier, Clifton, and Fischer 1985), we have documented a transient improvement in sensory function 12 hours after treatment. To observe early changes, we sometimes assessed these patients daily for three days after injury and then weekly for six weeks. Thus the timing of the examinations must be varied depending on the severity of injury and whether the treatments being tested are anticipated to produce rapid changes.

The role of neurophysiological testing

The neurological examination is often held as the golden standard against which all other tests of function have to be measured. However, neurological examinations have several drawbacks. First, assessments of sensory perception and motor strength depend on patient co-operation; unconscious patients are difficult to assess, for example. Second, phenomena such as latency and amplitude changes of responses within the central nervous system cannot be readily assessed. Third, there is an inherent subjectivity in the observations, as well as variability between examiners, which decrease the reliability of detecting small changes. Finally, rigorous neurological assessments of patients are time-consuming and expensive. Continuous monitoring is not feasible.

Neurophysiological testing of sensory and motor function can be used to assess neurological deficits. Somatosensory-evoked potentials (SSEPs), for example, have been routinely used in a number of spinal cord injury centres around the world (Young, 1982, 1985; Young and Berenstein 1985). Electromyographic studies can provide quantitative data concerning segmental reflexes, long loop pathways, and descending motor pathways (Bracken, Webb, and Wagner 1977). The number of investigators being trained in these techniques is increasing. The power of computerized instruments used for neurophysiological tests has grown exponentially in recent years, while prices of these instruments have steadily declined, making them more attractive. For frequent examinations of many patients, some of the neurophysiological tests are becoming more economical than repeated neurological examinations.

Properly applied evoked potential studies can yield reliable data concerning electrical manifestations of responses conducted through the spinal cord (Young 1985). Evoked potentials, however, test only the pathways being stimulated. For example, at NYU we activated cortical SSEPs from the median and posterior *tibialis* nerves. Results of these SSEP tests therefore pertain only to changes above the appropriate root entry zones. Also, the relationship of amplitude and latency of evoked potential components with spinal axonal conduction and number of axons is still far from clear. For example, cortical SSEPs in animal studies may recover to nearly normal amplitudes and latencies despite lesions of as much as 50 per cent of the spinal tracts. Functional activity usually transmits complex frequency and spatially coded bursts of spikes. Electrical stimulation used in neurophysiological tests delivers gross suprathreshold shocks to the peripheral or central nervous systems. An electrically activated burst of action potentials may conduct across spinal lesions that would strip critical information from more complex naturally coded signals. Finally, injured spinal axons often fatigue with the repetitive rapid shocks used to obtain averaged potentials, yielding false positive results.

Neurophysiological testing cannot replace neurological assessments. They can augment the information gathered, however, in several important ways. First, evoked potentials may demonstrate the presence of conduction in patients with no apparent function. For example, in 400 patients we have found that approximately 4–5 per cent of patients have definite SSEPs and no detectable sensory function in the appropriate dermatomes on admission. This percentage increases to 17 for chronic spinal injured patients. Second, neurophysiological tests can be carried out frequently in situations of rapid change. For example, evoked potentials can be obtained continuously from spinal injured patients in intensive care units or operating rooms. Third, neurophysiological tests are useful in unco-operative or unconscious patients. For example, we have utilized evoked potentials to determine whether or not spinal injury has occurred in comatose, malingering, or psychiatric patients. Fourth, certain evoked potential findings may give insights into causes of dysfunction. For example, latencies of evoked potentials may be useful for distinguishing between conduction defects and axonal loss.

Motor scoring

Based on the widely used system defined by the Nerve Injuries Committee (1943) of the Medical Research Council, muscle strength is often graded on a scale of 0–5. The criteria for the scores are given in Table 6.2. At NYU, we

Table 6.2. Criteria used in determining muscle strength scores (Nerve Injuries Committee 1943)

Score	Criteria
0	No contraction; i.e. total paralysis
1	Flicker or trace of contraction
2	Active movement, with gravity eliminated
3	Active movement against gravity
4	Active movement against gravity, and resistance
5	Normal power

Table 6.3. Segmental representation of muscles assessed

Label	Muscles examined	Segments[1]
Delta	Rhomboids, deltoids	C5 (C4, C6)
Forearm flexors	Biceps, *brachioradialis*	C5, C6
Hand flexors	Flexor *carpi radialis*	C6, C7 (C8)
Forearm extensors	Triceps	C7, C8
Hand extensors	Extensor *carpi ulnaris*	C7, (C8)
Opponens pollicis	*Opponens pollicis*	T1 (C8)
Upper abdomen and lower abdomen		T2–12
Hip flexors	Ilio-psoas, *sartorius*	L2, L3 (L1)
Hip extensors	*Gluteus maximus, medius*	L5, S1 (L4, S2)
Leg flexors	Hamstrings	L5, S1 (L4, S2)
Leg extensors	Quadriceps	L3, L4 (L2)
Foot flexors	*Tibialis* anterior	L4 (L5)
Foot extensors	Gastrocnemius	S2 (S1)
Extensor *hallus*	Extensor *hallucus longus*	L5 (S1)

[1] Segments with minor representations are given in parentheses.

score major muscle groups on each side, as listed in Table 6.3 (see also Fig. 6.1). These scores map the integrity of motor responses from the C4–S2 segmental levels. Note, however, the substantial redundancy in segmental representation. The upper and lower abdominal muscles are given scores of 0 and 1, indicating whether or not the patient can tense the muscle sufficiently for the muscle belly to be palpable. The muscle strength below the lesion level can be summed to reflect the general motor status of the patient.

Motor scores must be carefully interpreted for several reasons. First, the relationship of these arbitrary measures of muscle strength to spinal cord function may not be linear. For example, a score of 4 on an individual muscle may require recovery of 10-times more spinal axons than that needed to produce a score of 2. Second, the summing of muscle scores assumes that the muscle groups are equally represented in the spinal cord. For example, we have given greater weight to the lower limb muscles with fourteen muscle groups than to the upper limb muscles with twelve muscle groups. Third, misleading impressions can result from unevenly distributed scores. For example, recovery of six muscle groups to normality may have different functional implications to recovery of fifteen muscle groups to scores of 2, although they both produce identical total scores. Finally, muscle strength does not imply co-ordination.

The advantages of this scoring system, however, far outweigh the disadvantages. Most clinicians are well trained to apply this scale. The muscle

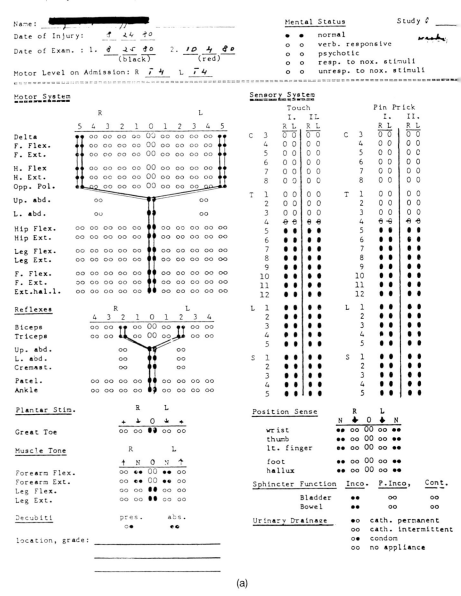

(a)

Fig. 6.1. The standard record of neurological examinations carried out on all spinal-injured patients admitted to the Bellevue Spinal Cord Injury Center at NYU. Developed by Dr B. Fischer, this form displays all the information gathered on two consecutive examinations and can be scanned at a glance. The data can be colour-coded for time after injury to facilitate comparison of changes. Most motor functions

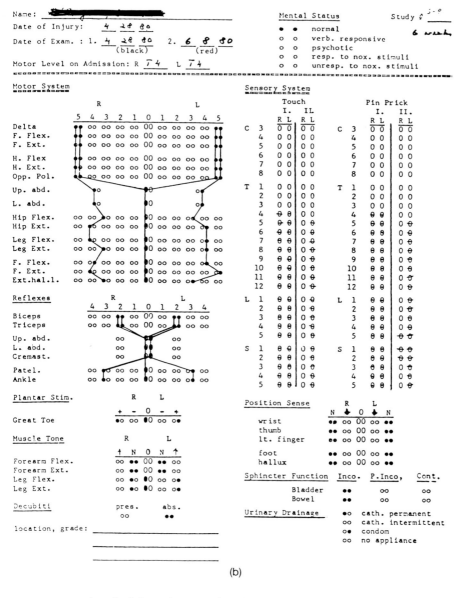

(b)

are represented on the left, and sensory functions on the right. The date of the injury and the dates of the examinations are listed at the top, along with lesion level and mental status; (a) represents records of a patient with severe T6 injury; (b) is from a patient with a partial T4 lesion.

groups chosen for assessment are relatively accessible and procedures for testing them are extensively published. Though some muscle groups give redundant information, the thirty-two muscle groups do represent reasonable outcome measures of treatment. Furthermore, subsets of the muscle groups can be analysed retrospectively. For example, strength recovery in extensor *hallucis longus* muscles may not contribute substantially to functional recovery, and represent L5–S1 segments which are already assessed by scores of the hip extensors and leg flexors; that particular score can simply be omitted from the analysis.

Sensory scoring

Sophisticated methods are available for quantifying many sensory modalities of the body. Exhaustive documentation of the sensory changes in spinal injured patients, however, is neither possible nor necessarily desirable. Relatively simple sensory examinations can yield the critical information needed for determining the lesion level and the extent of sensory loss below the lesion level. At NYU we chose to emphasize the spatial distribution of sensory changes rather than the detailed nature of the losses in any given dermatome. Sensory changes were scored as absent (0), abnormal (1), or normal (2) in every dermatome.

Like the motor scores, sensory scores must be cautiously interpreted. For example, recovery of normal sensation in half the dermatomes below the lesion level would yield the same score as partial recovery of all the dermatomes. The problem, however, is not as potentially misleading as in the case of muscle scores, where normal strength in one muscle is equivalent to partial recovery in 2–5 muscles. The use of fewer categories (0–2) eliminates the need for specialized equipment to characterize sensory changes, requiring only a pin. The large number of dermatomes tested and the assessment of both pin and touch modalities reduces the possibility that miscategorization of a few dermatomes would seriously bias results.

Documenting the distribution of sensory scores, dermatome by dermatome, is particularly useful for identifying lesion level. Since each dermatome represents one segmental level, the sensory level can easily be defined as the segment below which deficits are found. Although the lesion levels are relatively easy to identify in patients with severe spinal injuries, they are often ambiguous in patients with some preservation of function below the lesion. Discrepancies may also occur between motor and sensory levels. Patchy losses distal to the lesion site may occur. Given the overlap present in spinal segmental representation of muscles, determinations of lesion level from the motor findings is not as precise as with sensory testing of dermatomes. Change of lesion level is an important aspect of recovery that has often been neglected. In the cervical or lumbar area, improvement of motor function at several segmental levels carries major functional implications. Note, however, that transient fluctuations in sensory levels are common and may represent changes in the excitability of the spinal cord proximal to the lesion, with no long-term implications of improved sensory function.

Data analysis

In any statistical analysis of complex outcomes, such as in spinal cord injury, two approaches are possible. One is to use complex multivariate methods, which correlate many variables to yield a single outcome. The other is to distil the data into a one- or two-dimensional variable, which can then be assessed using simple statistical approaches. The former yields statistically rigorous evaluations of the data (Bracken, Collins, Freeman, Shepard, Wagner, Silten, Hellenbrand, Ransohoff, Hunt, Perot, Grossman, Green, Eisenberg, Rifkinson, Goodman, Meagher, Fischer, Clifton, Flamm, and Rawe, 1984; Bracken et al. 1985), but the process by which conclusions are reached is not always intuitively obvious and seldom yields insight into the nature of the outcome. The latter requires more manipulation of the data prior to analysis, but often gives a better intuitive grasp of the data. The best statistical analyses combine both approaches.

I will focus here briefly on the strengths and pitfalls of one method of distilling the data obtained with the NYU protocol. The easiest method of reducing complex motor or sensory scores into a single term is to sum them. Thus, for each patient, we can obtain a total sensory or motor score before and after treatment. Displayed in a scatter plot of points, with the initial scores represented on the abscissa and the follow-up scores on the ordinate, changes in large numbers of patients can be perceived in a glance. Patients showing no change between the initial and follow-up studies should fall on a line with a slope of one (unity slope line). Those who improve are on the left of the line and those who deteriorate are on the right. Figures 6.2 and 6.3 show the data for the two treated groups of patients. Group A received 1 g methylprednisolone sodium succinate (MPSS) and group B 100 mg MPSS within 48 hours of injury. Both groups were then treated daily for ten days.

With linear regression analysis, we can obtain the slope, ordinate intercept, correlation coefficient (r), and standard error for a line fitted to points by the least squares method. If r is significant, the standard error can be used to test whether the slope is significantly different from either unity slope or another line. Assume, for example, that the regression line lies to the left of and is significantly different from the unity slope line. A slope of less than one suggests that patients with lower admission scores improved more (i.e. in the lower left quadrant). A slope of one suggests that improvements occurred regardless of admission scores. A slope of more than one suggests

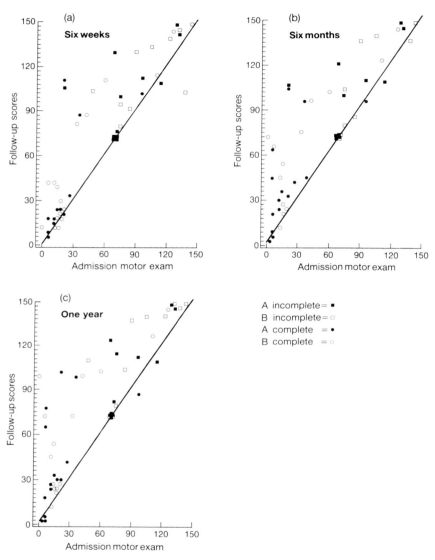

Fig. 6.2. Scatter plot of motor scores, comparing admission with follow-up scores. The motor scores represent sums of muscle strength grades (0–5) for thirty muscle groups, for a possible total of 150. For each point, admission scores are represented on the abscissa and follow-up scores on the ordinate. The times of the follow-up examinations are indicated as six weeks (a), six months (b), or one year (c). The individual points are segregated as belonging to patients with 'complete' (filled) and 'incomplete' (open) lesions on admission, for those treated with 1 g (A) and 100 mg (B) of methylprednisolone sodium succinate given intravenously within 48 hours after injury and then daily for ten days.

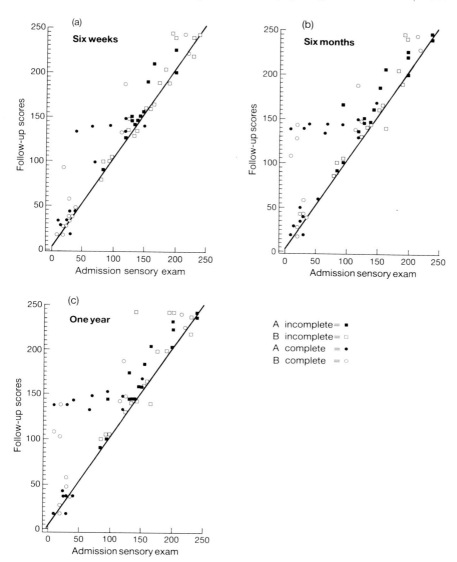

Fig. 6.3. Scatter plot of sensory scores, comparing admission with follow-up scores. The sensory scores represent the sum of pinprick, touch, proprioception, and deep pressure scores. The pinprick scores (0–2) and the touch scores (0–2) were individually obtained from each of fifty-six dermatomes tested. The proprioceptive joint scores (0–2) were obtained from ten joints. The deep pressure scores (0–2) were scored for each leg. These add for a maximum sensory score of 248. See legend of Fig. 6.2 for a fuller explanation.

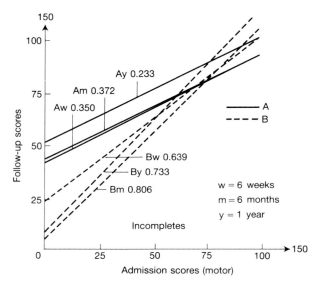

Fig. 6.4. Linear regression analysis of motor scores in incomplete patients. The regression lines were calculated from data shown in Fig. 6.2 for patients with some preservation of sensory function below the lesion level on admission ('incomplete'). Six lines are shown. Solid lines represent data from patients treated with 1 g MPSS (A) and dashed lines from data patients treated with 0.1 g MPSS (B) within 48 hours after injury and then daily for ten days. In each line label, the letters w, m, and y indicate, respectively, data obtained six weeks, six months, and one year after injury. The numbers indicate the correlation coefficient for the lines.

that the improvement was most marked in patients with higher admission scores (i.e. in the upper left quadrant). Figure 6.4 shows an example of such an analysis for motor scores in patients with some motor preservation below the lesion level. The slope of line Am 0.372 (1 g MPSS at six months) is significantly different from that of line Bm 0.806 (100 mg MPSS at six months), with $p < 0.05$. Due perhaps to the smaller number of patients at one year, the slopes of lines Ay 0.233 (1 g MPSS at one year) and By 0.733 (100 mg MPSS at one year) are not significantly different.

Total scores do not distinguish between changes in lesion level and more distal recovery. The lesion level also influences the total score. For example, the same total score may represent a partial proximal lesion or a more severe distal lesion. Nevertheless, significant changes of total score between initial and follow-up examinations are indicative of a treatment effect regardless of the lesion level. Specific scores can be always be analysed separately to ascertain the location of the effect. Total scores represent a reasonable and sensitive method of screening treatment effects.

Towards a standard protocol for evaluating spinal cord injury

As new treatments are found that produce functional recovery in animal spinal cord injury models, the need for clinical trials will increase. Given the diversity of clinical spinal cord injury and the multiplicity of pathological mechanisms believed to influence recovery from spinal cord injury, it is unlikely that any single treatment will yield dramatic results that can be detected in a small group of randomly selected patients. Clinical trials must assess large numbers of patients, which necessarily involves the joint efforts of many institutions. If a treatment is found to have modest beneficial effects, more clinical studies must be carried out to determine optimal dosage or efficacy of combined therapies. The treatment effects in such studies will be even more subtle. Therefore the development of better evaluation protocols that can detect small changes with greater sensitivity and reliability is essential to a successful outcome in the quest for effective treatments of spinal cord injury.

The experience at NYU demonstrates the feasibility of more detailed and frequent standardized neurological examinations. Since a major study (NASCIS) involving more than twelve spinal injury centres in the USA is based on a subset of the NYU protocol, future standard protocols would benefit from compatibility so that data bases can be shared. It is important to emphasize, however, that although the NYU and NASCIS evaluation protocols are more comprehensive than any utilized before, they do not approach the sensitivity and reliability required in anticipated studies. For example, large numbers of patients are needed to show significant results; follow-up periods of a year or more are required; the protocols contain gaps and redundancies. Neurophysiological tests are still not being systematically applied in most centres participating in NASCIS.

Efficient evaluation protocols must be developed that not only can reliably detect small therapeutic effects but that can also make do with fewer patients and shorter follow-up periods to show the therapeutic effects. Already the limited number of acutely injured patients available for clinical study has severely restricted the number of treatment paradigms that can be tested in current trials. For example, to randomize 300 acutely injured patients for a three-arm clinical trial of naloxone, methylprednisolone, and placebo, NASCIS has to monopolize the resources of twelve or more spinal injury centres for three years. The typical 1–2-year period of follow-up required to determine outcome also contributes to the cost and length of clinical trials. If reliable predictions of recovery can be made within six weeks after injury, trials can be greatly shortened. Finally, examination protocols have to be developed for chronic spinal cord injury. Thus much work remains to be done.

Summary

Recent treatment studies in spinal cord injury have tended towards more detailed and frequent examinations. To document small therapeutic effects, careful motor and sensory examinations are needed. The spatial distribution of deficits can be used to specify the lesion level. Scoring of sensory changes, dermatome by dermatome, and strength in representative muscle groups provides quantitative measures of lesion severity. Frequent examinations can yield the time course of recovery. Given the diversity of clinical spinal injury and the multiplicity of pathological mechanisms believed to influence recovery, it is unlikely that any single treatment will dramatically improve recovery in small numbers of randomized patients. Current clinical trials are limited by the large number of patients that must be tested and the long follow-up periods required. Success in the quest for effective treatments of spinal cord injury will depend on the use of efficient evaluation protocols able to detect modest therapeutic effects in fewer patients and over shorter follow-up periods.

Acknowledgements

I thank my colleagues in the Department of Neurosurgery at NYU for their help, particularly Boguslav Fischer who examines and scores our spinal injured patients and Donna Whittam for her assistance as the Clinical Co-ordinator of the Spinal Cord Injury Center at NYU. This work is supported by grants from NIH (NS10164, NS15990).

References

Bracken, M. B. and Collins, W. F. (1985). Randomized clinical trials of spinal cord injury treatment. In *Central nervous system trauma status report* (ed. D. P. Becker and J. T. Povlishock), pp. 303–12. NIH NINCDS, Washington, DC.

Bracken, M. B., Webb, S. D., and Wagner, F. C. (1977). Classification of the severity of acute spinal cord injury: implications for management. *Paraplegia* **15**, 319–26.

Bracken, M B., Shepard, M. J., Hellenbrand, K. G., Collins, W. F., Leo, L. S., Freeman, D. F., Wagner, F. W., Flamm, E. S., Eisenberg, H. M., Goodman, J. H., Perot, P. L., Green, B. A., Grossman, R. G., Meagher, J. N., Young, W., Fischer, B., Clifton, G. L., Hunt, W. E., and Rifkinson, N. (1985). Methylprednisolone and neurological function one year after injury: results of the National Acute Spinal Cord Injury Study. *J. Neurosurg.* **63**, 704–13.

Bracken, M. B., Collins, W. F., Freeman, D. F., Shepard, M. J., Wagner, F. W., Silten, R. M., Hellenbrand, K. G., Ransohoff, J., Hunt, W. E., Perot, P. L., Grossman, R. G., Green, B. A., Eisenberg, H. M., Rifkinson, N., Goodman, J. H., Meagher, J. N., Fischer, B., Clifton, G. L., Flamm, E. S., and Rawe, S. E.

(1984). Efficacy of methylprednisolone in acute spinal cord injury. *J. Am. Med. Ass.* **251**, 45–52.

Chehrazi, B., Wagner, F. C., Collins, W. F., and Freeman, D. H. (1981). A scale for evaluation of spinal cord injury. *J. Neurosurg.* **54**, 310–15.

Cheshire, D. J., (1970). A classification of the functional end-results of injury to the cervical spinal cord. *Paraplegia* **8**, 70.

Ducker, T. B., and Wallek, C. A. (1985). Recovery from cord injury. In *Central nervous system trauma status report* (ed. D. P. Becker and J. T. Povlishock), pp. 369–74. NIH NINCDS, Washington, DC.

Flamm, E. S., Young, W., Collins, W. F., Piepmeier, J., Clifton, G. L., and Fischer, B. (1985). A phase I trial of naloxone treatment in acute spinal cord injury. *J. Neurosurg.* **63**, 390–7.

Frankel, H. L., Hancock, D. O., Hyslop, G., Melzak, J., Michaelis, L. S., Ungar, G. H., Vernon, J. D. S., and Walsh, J. J. (1967). The value of postural reduction in the initial management of closed injuries of the spine with paraplegia and tetraplegia. *Paraplegia* **7**, 179–92.

Guttman, L. (1973). *Spinal cord injuries—comprehensive management and research* (2nd edition). Blackwell Scientific Publications, Oxford.

Lucas, J. T. and Ducker, T. B. (1979). Motor classification of spinal cord injuries with mobility, morbidity, and recovery indices. *Am. Surgeon* **45**, 151–8.

Maynard, F. M., Reynolds, G. G., Fountain, S., Wilmot, C., and Hamilton, R. (1979). Neurological prognosis after traumatic quadriplegia: regional spinal cord injury system. *J. Neurosurg.* **50**, 611–16.

Nerve Injuries Committee (1943). *War memorandum No. 7* (2nd edition). Medical Research Council, London.

Young, W. (1982). Correlation of somatosensory evoked potentials and neurological findings in spinal cord injury. In *Early management of acute spinal cord injury* (ed. C. H. Tator), pp, 153–65. Raven Press, New York.

Young, W. (1985). Blood flow, metabolic and neurophysiological mechanisms in spinal cord injury. In *Central nervous system trauma status report* (ed. D. P. Becker and J. T. Povlishock), pp. 463–75. NIH NINCDS, Washington DC.

Young, W. and Berenstein, A. (1985). Somatosensory evoked potential monitoring of intraoperative procedures. In *Spinal cord monitoring* (ed. J. Schramm), pp. 197–203. Springer, Berlin.

Neurological and medical assessment of function and dysfunction

Alain B. Rossier

Introduction, or a word of caution

Any conscious or unconscious individual involved in a major injury, e.g. car collision or fall from a height, who is brought to or even walks into an emergency room should be considered, until further disproven, as having sustained major spinal damage and minimal cord involvement. The rule of careful handling of the patient during a thorough evaluation of sensory–motor functions and reflexes should be particularly adhered to during the radiographic examination of the entire spine, including views of the odontoid. This is the best way to prevent patients without initial neurological deficit from becoming tetraplegics because of inadequate X-ray imaging, especially of the cervicothoracic region, or because of inappropriate manipulations during radiological investigations (Rogers 1957; White and Moss 1978; Scher 1981). The need for X-rays of the entire spine should be emphasized because more than one or two contiguous spinal levels may be involved. A review of 146 cervical and fifty-four thoracic and thoracolumbar injuries has revealed seven dorsal concomitant and one cervical fracture, respectively.

Initial evaluation

If the patient is conscious and does not complain of sensory or motor loss but displays asymmetry of deep tendon reflexes he should be approached with a high degree of suspicion. A careful neurological evaluation may show this asymmetry to be accompanied by discrete sensory changes—hypalgesia, hypoaesthesia—the upper level of which will be seen to correspond to a metameric segment.

In the unconscious patient the presence and absence of reflexes in the upper and lower extremities points, respectively, at a thoracic or dorsolumbar injury. Areflexia in the four extremities and evidence of paradoxical breathing—exclusive of flail chest—are indicative of cervical cord damage. Conversely, paradoxical respiration with positive deep tendon reflexes in the upper extremities is suggestive of a high dorsal injury (Sandor 1966).

Table 7.1. Distribution of causes of 300 consecutive acute spinal cord injuries admitted[1] to the Spinal Cord Injury Service of the West Roxbury Veterans' Administration Medical Centre

Cause	Percentage occurrence
Motor vehicles	45
Falls	24
Acts of violence	15[2]
Water sports	12
Other sports	3

[1] Of these patients 10 were admitted within 4 hours of injury, 12 within 4–6 hours of injury, 16 within 6–8 hours of injury, and 17 within 8–12 hours of injury.
[2] Half of these were gunshot wounds.

About 50 per cent of spinal cord injuries arise from causes other than road traffic accidents (Table 7.1). A review of 300 consecutive acute spinal cord injuries admitted to the Spinal Cord Injury Service of the West Roxbury Veterans' Administration Medical Center (Boston, USA) over a seven-year period is outlined in Table 7.1. Since 25 per cent of these cases were not veterans, but civilians, this sampling can be considered as rather representative of the US population. There were 188 tetraplegics (63 per cent) and 112 paraplegics (37 per cent). One hundred and one patients were admitted within 24 hours of injury and fifty-seven within 2–3 days post-injury.

Road traffic accidents are often indicative of major violence, especially when the naturally 'well-splinted' thoracic spine is involved. It follows that frequent intrathoracic complications, such as haemopneumothorax or, more rarely, aortic damage, have to be taken into consideration, whereas at times dorso-lumbar and lumbar fractures will be found to be associated with intra-abdominal lesions (Meinecke 1968). An open minilap lavage should be carried out in any doubtful situation because abdominal injuries are difficult to diagnose in acute spinal cord injuries when there is loss of abdominal pain sensation and other defence mechanisms.

In contrast to signs presented in surgical shock, the blood pressure in cervical and dorsal injuries is usually low (90–100 mm Hg), but associated with bradycardia and a pulse of good volume. Features of surgical shock should therefore raise a high degree of suspicion of internal bleeding (Hardy and Rossier 1975). Surgical shock should be differentiated from spinal shock, although both conditions may exist. Spinal shock, as first defined by Hall (1843) and later by Sherrington (1899) serves no useful purpose other than to describe a state of profound depression of reflex activity below the level of an acute spinal cord injury. It bears no relationship to the completeness or incompleteness of the neurological lesion, although more severe cord damage appears to be accompanied by a period of spinal shock of a more profound degree and of a prolonged duration.

In contradistinction to some statements (Stauffer 1976), the presence or absence of reflex activity at the level of the *conus medullaris* (anal wink, bulbocavernosus reflex, anal tone) does *not* bear any direct relationship to the completeness or incompleteness of the spinal cord lesion. In the majority of cases—exclusive of *conus* destruction and complete *cauda equina* damage—all or some of the *conus* reflexes will be present (Rossier, Fam, di Benedetto, and Sarkarati 1980), whether the lesion is complete or incomplete.

Preservation of some sensation in the sacral segments—sacral sparing—should be carefully looked for. It may be the only sign of incompleteness of the lesion, pointing to a possible neurological recovery. It has been shown that motor-complete patients who have some preservation of pain sensation, as minimal as unilateral sparing, have a better prognosis for neurological recovery than those with some preservation of touch only (Foo, Subrahmanyan, and Rossier 1981).

Somatosensory-evoked potentials (SSEPs)

The absence of cortical SSEPs in patients with acute spinal cord injuries has been reported as indicative of severe cord damage and of a poor prognosis for recovery. In patients with marked sensory and motor deficits SSEPs have been shown to be of prognostic value as they preceded major clinical recovery at a time when the patient displayed only discrete signs of improvement (Rowed, McLean, and Tator 1976). While the absence of SSEPs has been shown to be associated with no clinical recovery, the presence of SSEPs has been of little value in predicting the potential recovery (York, Watts, Raffensberger, Spagnolia, and Joyce 1983). In spite of our as yet limited experience with SSEPs in acute cord injuries, they do not appear to be of much greater diagnostic and prognostic value than regular careful neurological examinations carried out at close intervals.

Spinal cord hypothermia

Experimental conflicting data have been reported in regard to the protective value of localized cooling upon the traumatized cord: reduction of spinal cord oedema, decrease of intramedullary haemorrhage, reduction of metabolic demand, improvement of local blood perfusion, removal of noxious catabolic agents (Albin, White, Locke, Massopust, and Kretchmer 1967; Albin, White, Acosta-Rua, and Yashon 1968; Tator and Deecke 1972;

Thienprasit, Bantli, Bloedel, and Chou 1975; Tator 1976). Data on randomized, standardized clinical studies are as yet unavailable. Our present clinical knowledge is based on a limited number of reports, the conclusions of which are ambiguous and unconvincing (Koons, Gildenberg, Dohn, and Henoch 1972; Meacham and McPherson 1973). The results obtained by Bricolo, Ore, da Pian, and Faccioli (1976) from eight spinal cord injury patients are inconclusive, mainly because insufficient data were collected on the neurological status before treatment of the patients who 'recovered' after cooling. Conversely, the marked neurological improvement observed by Hansebout, Tanner, and Romero-Sierra (1984) in 48 per cent of fifty-two complete spinal cord injury patients cooled within eight hours post-injury would appear to justify further trials with this technique, provided it could be initiated within four hours of injury. Similar views have been expressed by White (1986).

It has been shown that acute cord lesions progressively develop time-dependent characteristics which involve at various degrees the vascularization of both the grey and the white matter. A 1–4-hour cut-out time seems to be the limit for possible therapeutic effects (Assenmacher and Ducker 1971; Dohrmann and Allen 1975; Dohrmann, Wagner, and Bucy 1971). Unhappily there are but a few acute patients to be admitted in a spinal cord injury service within 4–6 hours of injury (see Table 7.1). One also has to consider that laminectomy which is a prerequisite of any hypothermic procedure has but a few therapeutic indications *per se*. In addition, it increases morbidity (Bohlman 1979) and further destabilizes the damaged spine. In view of these limitations and of the lack of clinical evidence that hypothermia is beneficial in major cord injuries, this technique should not be recommended as a proven treatment of acute cord injuries until further experimental data and more clinical experience has been gained.

Steroids

In spite of enthusiastic reports on the benefits of steroids in spinal cord injuries (Bucy 1973) there have been but two recent controlled studies of their effects in these patients (Bracken, Collins, Freeman, Shepard, Wagner, Silten, Hellenbrand, Ransohoff, Hunt, Perot, Grossman, Green, Eisenberg, Rifkinson, Goodman, Meagher, Fischer, Clifton, Flamm, and Rawe 1984; Bracken, Shepard, Hellenbrand, Collins, Leo, Freeman, Wagner, Flamm, Eisenberg, Goodman, Perot, Green, Grossman, Meagher, Young, Fischer, Clifton, Hunt, and Rifkinson 1985). Experimental studies have postulated that steroids stabilize membranes and reduce ischaemia and oedema, and the groups treated with steroids showed some benefit from treatment when compared to the untreated control groups (Ducker and Hamit 1969; Green, Kahn, and Klose 1980; Braughler and Hall

1983). Other experimental studies have shown that animals treated with dexamethasone or methylprednisolone show significantly better recovery than do sham controls (Lewin, Pappius, and Hansebout 1972; Hall, Wolf, and Braughler 1984). It follows that the role of cord oedema in potentially reversible cord lesions remains as yet unanswered. Other clinical investigators have concluded that, although steroids do influence post-wounding haemorrhagic cord necrosis to a limited extent, severe spinal cord injury patients do not do much better with steroids (Osterholm 1974). Corroboratively, some extensive clinical studies have shown that steroids not only did not improve neurological recovery, but that their use was associated with gastrointestinal bleeding (Bohlman 1979).

More recent clinical and experimental reports on the possible benefit of initial higher doses of steroids in acute spinal cord injuries have cast doubts as to their therapeutic value (Gunby 1982; Faden, Jacobs, Patrick, and Smith 1984), although strong evidence pointing at the inefficiency of high-dose versus standard-dose methylprednisolone has been presented (Bracken et al. 1985).

Our own experience in more than 400 acute cases admitted within a few days of injury points at the lack of efficacy of steroids in complete lesions. However, in incomplete lesions steroids at times appear to be of definite benefit, as shown by the following example.

A 26-year-old patient was admitted with fractures of C4 and C5 and a sensory–motor very incomplete tetraplegia C4–5 bilaterally with nearly complete monoplegia of the left arm. After an initial loading dose of 125 mg solumedrol, the patient was placed on 5 mg dexamethasone every six hours, the dose tapering over the next eight days. The initially progressive improvement in the function of the left-arm muscles over that same period of time came suddenly to a halt and was followed by a brisk motor deterioration twenty-four hours after discontinuation of steroids. A metrizamide myelogram did not disclose any cord compression, but showed at the C3–4 level a markedly enlarged cord which was interpreted as secondary to oedema. At completion of the myelogram, intravenous dexamethasone was started immediately. Within two hours of treatment, a dramatic improvement of muscle function in the left upper extremity was reported by the patient and could also be demonstrated objectively. Muscle function continued to improve until discharge one month later, at which time all muscles were graded 3+ (out of 5). The succession of events in that particular case seems to point at a cause-and-effect relationship between treatment with steroids and neurological evolution. Although one swallow does not make a summer, it has been our policy to give steroids in incomplete lesions over a 10–12-day period post-injury. Whether steroids are of any benefit in the treatment of complete lesions still remains unanswered as yet.

It may well be that treatment with naloxone or thyrotropin-releasing hormone will be shown to be the medication of choice in promoting better neurological recovery (Arias 1985). Faden, Jacobs, Smith, and Holaday (1983) have presented experimental evidence that these pharmacologic opiate antagonists are more efficient than dexamethasone in improving neurological function of acutely injured cats. The mode of action of these drugs is as yet still conjectural, although it has been suggested that they could act in altering local metabolism and microvascularization. It should still be added that although the role of steroids in the genesis of 'stress ulcers' has not yet been demonstrated, acute spinal cord injury patients who are known to be prone to developing gastrointestinal ulcers may represent a higher-risk population (Epstein, Hood, and Ransohoff 1981).

Gastrointestinal complications

During the acute post-traumatic period, which is closely related to spinal shock most of the time, there is a paralytic ileus with absent or poor bowel sounds. This condition may result in acute dilatation of the stomach, with nausea, vomiting, and aspiration pneumonia. Moreover, the gastric dilatation pushes up the diaphragm and further compromises an already borderline pulmonary function, as can be seen in tetraplegia and high dorsal paraplegia without abdominal and intercostal muscle functions. Such an unrecognized complication has been seen to be lethal in some instances. Most of the time intravenous fluids and immediate bowel retraining with daily suppository and digital stimulation of the rectoanal region will overcome this condition. However, in these few cases of acute abdominal distension it is important to start nasogastric suction in order to prevent vomiting and aspiration of vomitus which is facilitated by the supine position and cough impairment (paralysis of abdominal muscles). Under these circumstances fluid losses will have to be replaced.

Another serious and often deadly complication in the acute stage of injury is the so-called 'stress ulcer'. Since the acute patient in spinal shock presents with a flaccid paralysis of his abdominal muscles and absent bowel sounds (paralytic ileus), he will be unable to display any of the usual signs which may accompany a gastrointestinal perforation in a chronic spinal cord injury patient, such as unilateral increased tone of abdominal muscles or increased sweating (autonomic hyperreflexia). In complete cervical or mid-dorsal lesions subjective symptoms will be absent. The diagnosis of a perforated ulcer, therefore, may be difficult. The physician will have to rely upon signs such as tachycardia, hypotension, drop in the haematocrit, pneumo-peritoneum, and presence of blood in the stools or in the gastric contents. A rather reliable and frequent clinical sign indicating an impending gastrointestinal perforation lies in the patient's complaint of referred pain in the shoulder tip.

Acute spinal cord injuries, not infrequently, are complicated by gastrointestinal bleeding and death (Schneider 1955; Kewalramani 1979; El Masri, Cochrane, and Silver 1982). Patients with complete cervical cord lesion would appear to be at greater risk than other groups of patients (Soderstrom and Ducker 1985). Such factors as hypoxia, hypoxaemia, and/or modifications in gastric secretion may play a role in the genesis of stress ulcers. Early recognition of this potentially lethal complication is, therefore, of the essence. It is worth mentioning that beside adequate ventilation, cimetidine may be of use in decreasing the incidence of gastrointestinal stress ulcers, although the findings reported in a recent study of critically ill patients have pointed at better results with antacids than with cimetidine in preventing acute gastrointestinal bleeding (Priebe, Skillman, Bushnell, Long, and Silen 1980). However, such an antacid regimen is not appropriate in acute spinal cord injury patients who present initially with paralytic ileus which may last several days. Diarrhoea due to antacids in patients without sphincter control may cause problems (Priebe *et al.* 1980). Therefore, during the period of paralytic ileus we routinely give 300 mg cimetidine intravenously every six hours to all of our acute patients, before reverting to oral therapy as soon as the bowel has resumed its activity. This treatment is prescribed for 3–4 weeks from the day of injury, watching the blood formula since this medication has been known to cause granulocytopenia. Although animal experimental data (Levine, Schwesinger, Sirinck, Jones, and Pruitt 1978) and initial clinical investigations on the prevention of stress ulcers after severe head trauma look encouraging (Halloran, Zfass, Gayle, Wheeler, and Miller 1980), further studies are indicated to evaluate the effect of this medication in the prevention of stress ulcers in the acute spinal cord injured. Preliminary data with this medication in our acute patients seem to point at a decrease of severe gastrointestinal bleeding over a five-year period during which we initiated this prophylactic regimen. Whether ranitidine should replace oral cimetidine for the oral prophylactic regimen after the initial intravenous therapy remains conjectural at this time (McCarthy 1983).

Deep venous thrombosis and pulmonary embolism

Deep venous thrombosis (DVT) and pulmonary embolism (PE) represent two frequent serious complications in acute spinal cord injured for whom prophylactic anticoagulant therapy has been recommended as early as 1964 in a survey of thirty-two acute patients, five (15.6 per cent) of whom died following PE which was demonstrated at autopsy (Rossier and Brunner 1964).

Reported frequency of fatal PE in the acute spinal cord injured has been from 15 to 48 per cent (Tribe 1963; Rossier and Brunner 1964; Walsh and Tribe 1965; Wolman 1965; Watson 1978). This frightening complication is, therefore, not at all infrequent. By the same token, the occurrence of DVT in untreated patients has been shown to be high, 14–25 per cent (Silver 1974; Perkash, Prakash, and Perkash 1978; Watson 1978). The majority of DVT occurs within the first four weeks of injury and our experience in 194 acute cases concurs with that finding. It must also be underlined that 45 per cent of PE takes place at the same time as DVT, and half of them are instantly fatal (Watson 1968). From 1978 to 1983, thirty-five cases of DVT with PE were diagnosed in 194 acute spinal cord injury patients. One-third of these thirty-five cases had the DVT–PE occurring during the third and fourth week post-injury.

In acute spinal cord injuries prophylactic anticoagulant therapy with subcutaneous heparin or sodium warfarin has been shown to be efficient in decreasing the incidence of DVT (Hachen 1974; El Masri and Silver 1981). Unless contraindicated because of such reasons as skull fracture or tracheostomy, it is our policy to initiate treatment from the third day of injury with subcutaneous heparin, 5000 units three times a day, checking the platelet count twice a week since heparin may cause thrombocytopenia. Our preliminary results over a five-year period of time point at a decreased incidence of DVT and PE in our acute patients on heparin (Table 7.2).

Since it has been shown that there are two peaks of hypercoagubility, one at the end of the first month, the second at the end of the third month (Dollfus, Bouchier, Henon, Jung, and Issler 1980), it has been our policy to continue anticoagulant therapy for three months, whether or not the patient is in a wheel-chair. Premature discontinuation of treatment before that time has resulted in DVT in a few cases.

We do not think that it is advisable to start an anticoagulant prophylactic therapy earlier than three days post-injury. It has been shown that four hours after an acute trauma was applied to the spinal cord there were haemorrhages in the central grey matter and in the subarachnoid space (Wagner, van Gilder, and Dorhmann 1978). Premature anticoagulation may increase, therefore, the extent of early intramedullary bleeding, also documented by Wolman (1965) in his pathological studies of the spinal cord

Table 7.2. Deep venous thrombosis (DVT) and pulmonary embolism (PE) in 166 acute spinal cord injuries

Treatment	Number of cases	DVT	DVT + PE	Total[1]
No anticoagulant	62	10	4	14 (23)
Heparin t.i.d.	104	10	4	14 (13.5)[2]

[1] Percentage values in parentheses.
[2] Adjusted [6(6)] after exclusion of cases improperly treated.

of patients deceased within a few days of injury. On the other hand, DVT does not seem to appear within the first 2–3 days post-injury.

Our regimen of low-dose heparin has not resulted in any visible bleeding in joints or in genitourinary or gastrointestinal tracts. With regard to the latter, we agree with Epstein *et al.* (1981) that the incidence and degree of gastrointestinal bleeding did not seem to be affected by prophylactic low-dose heparin. Out of a series of forty-nine acute spinal cord injury patients who were admitted to our service and who were placed on low-dose heparin, five (10 per cent) experienced minor gastrointestinal bleeding which did not require blood transfusion. A sixth patient was admitted fifty-nine hours post-injury with gastrointestinal bleeding which required four units of blood over a period of one week. At the time of the bleeding, the patient was not on heparin. A bulbar ulcer was diagnosed and treated conservatively. In comparison with the heparin group, there were twenty-five acute spinal cord injuries who were not placed on heparin. Two developed minor gastrointestinal bleeding; i.e. an incidence of 8 per cent.

Respiratory problems

Respiratory therapy every 2–3 hours is of great importance because of the known high rate of deaths due to respiratory failure in acute tetraplegics. It has been shown that in acute tetraplegics during the period of spinal shock the EMG activity of accessory respiratory muscles such as the sternomastoids was of small amplitude requiring the maximum breathing efforts on the part of the patient. Action potentials from the intercostal muscles in response to the act of breathing were practically absent (Guttmann and Silver 1965). The vital capacity was less than 2 l and in cervical and high dorsal injuries the forced vital capacity was 32 per cent and 43 per cent of predicted values, respectively (Guttmann and Silver 1965; Ohry, Molho, and Rozin 1975). Therefore, in acute high cord injuries the accessory respiratory muscles contribute little to tidal ventilation, and respiratory therapy every 2–3 hours may be of invaluable help in decreasing pulmonary complications.

Patients with cervical and high dorsal lesions with associated small vital capacities and an inability to cough properly are candidates for atelectasis and decreased arterial PO_2, even when the remaining limited ventilation is sufficient to maintain a normal carbon dioxide tension. Accumulation of secretions within the lungs and atelectasis prevent effective oxygenation of the blood and elimination of carbon dioxide. The resulting hypercapnia associated with a low alveolar PO_2 leads to intensive pulmonary vasoconstriction which may be life threatening.

When the deficit of the respiratory motor function is severe enough to result in inadequate ventilation and hypercapnoea, insertion of an orotracheal tube with appropriate respiratory support is indicated. Should it fail to bring the situation back under control within 2–3 weeks, consideration should be given to a tracheostomy. It has been our experience that endotracheal tubes with low-pressure cuffs deflated every two hours for 10–15 minutes have been well tolerated over that period of time, which allowed several tetraplegics to be eventually weaned off the respirator.

Because of the aforementioned cardiopulmonary complications, it is necessary to assess the respiratory rate, the tidal volume, and arterial blood gases in order to detect early rising serum bicarbonate and arterial carbon dioxide tension pointing at the deterioration of respiratory function.

Due to vasomotor paralysis the vascular space is already expanded and, thus, pulmonary oedema is a pending complication of fluid overload. However, it only occurs when the volume of intravenous fluids exceeds the amount necessary to maintain a cardiac pre-load sufficient to accommodate the increased vascular compliance. It follows that measurements of the hourly urine output and of its concentration by osmolarity are indicated.

In some patients, particularly those who are breathing with a small tidal volume, tracheal suction as well as the change of tracheal cannula may be accompanied by bradycardia and even cardiac arrest (Dollfus and Frankel 1965; Frankel, Mathias, and Spalding 1975; Welply, Mathias, and Frankel 1975). Schneider (1955) has reported a fatal outcome in a tetraplegic patient who ceased breathing following suction from the nasopharynx. The explanation lies in a vagovagal reflex, the action of which is unopposed by sympathetic activity or by the pulmonary (inflation) vagal reflex. Hypoxia adds to this condition since it is known that stimulation of arterial receptors by hypoxia is accompanied by reflex bradycardia. Since hypoxia may be caused by suctioning a patient previously normoxic, it seems advisable to give oxygen to all patients prior to the manœuvre. Subcutaneous, intramuscular, or intravenous atropine (0.4–0.6 mg) abolishes the vagovagal reflex and may be used either therapeutically or prophylactically.

References

Albin, M. S., White, R. J., Acosta-Rua, G., and Yashon, D. (1968). Study of functional recovery produced by delayed localized cooling after spinal cord injury in primates. *J. Neurosurg.* **29**, 113–20.

Albin, M. S., White, R. W., Locke, G. S., Massopust, L. C., and Kretchmer, H. E. (1967). Localized spinal cord hypothermia. Anesthetic effects and application to spinal cord injury. *Anaesth. Analg.* **46**, 8–16.

Arias, M. J. (1985). Effect of naloxone on functional recovery after experimental spinal cord injury in the rat. *Surg. Neurol.* **23**, 440–2.

Assenmacher, D. R. and Ducker, T. B. (1971). Experimental traumatic paraplegia. The vascular and pathological changes seen in reversible and irreversible spinal cord lesions. *J. Bone Joint Surg.* **53**(A), 671–80.

Bohlman, H. H. (1979). Acute fractures and dislocations of the cervical spine. An analysis of three hundred hospitalized patients and review of the literature. *J. Bone Joint Surg.* **61**(A), 1119–42.

Bracken, M. B., Collins, W. F., Freeman, D. F., Shepard, M. J., Wagner, F. W., Silten, R. M., Hellenbrand, K. G., Ransohoff, J., Hunt, W. E., Perot, P. L., Grossman, R. G., Green, B. A., Eisenberg, H. M., Rifkinson, N., Goodman, J. H., Meagher, J. N., Fischer, B., Clifton, G. L., Flamm, E. S. and Rawe, S. E. (1984). Efficacy of methylprednisolone in acute spinal cord injury. *J. Am. Med. Ass.* **251**, 45–52.

Bracken, M. B., Shepard, M. J., Hellenbrand, K. G., Collins, W. F., Leo, L. S., Freeman, D. F., Wagner, F. C., Flamm, E. S., Eisenberg, H. M., Goodman, J. H., Perot, P. L., Green, B. A., Grossman, R. G., Meagher, J. N., Young, W., Fischer, B., Clifton, G. L., Hunt, W. E., and Rifkinson, N. (1985). Methylprednisolone and neurological function 1 year after spinal cord injury. Results of the National Acute Spinal Cord Injury Study. *J. Neurosurg.* **63**, 704–13.

Braughler, J. M. and Hall, E. D. (1983). Lactate and pyruvate metabolism in injured cat spinal cord before and after a single large intravenous dose of methylprednisolone. *J. Neurosurg.* **59**, 256–61.

Bricolo, A., Ore, G. D., da Pian, R., and Faccioli, F. (1976). Local cooling in spinal cord injury. *Surg. Neurol.* **6**, 101–6.

Bucy, P. C. (1973). Editorial. Emergency treatment of spinal cord injury. *Surg. Neurol.* **1**, 216.

Dorhmann, G. J. and Allen, W. E. (1975). Microcirculation of traumatized spinal cord. A correlation of microangiography and blood flow patterns in transitory and permanent paraplegia. *J. Trauma* **15**, 1003–13.

Dohrmann, G. J., Wagner, F. C., and Bucy, P. C. (1971). The microvasculature in transitory traumatic paraplegia. *J. Neurosurg.* **35**, 263–71.

Dollfus, P. and Frankel, H. L. (1965). Cardiovascular reflexes in tracheostomized tetraplegics. *Paraplegia* **2**, 227–31.

Dollfus, P., Bouchier, J. J., Henon, P., Jung, G., and Issler, M. (1980). The value of the thromboelastogram as means of monitoring the sub-cutaneous preventive action of calcium heparinate. *Paraplegia* **18**, 157–66.

Ducker, T. B. and Hamit, H. F. (1969). Experimental treatments of acute spinal cord injury. *J. Neurosurg.* **30**, 693–7.

El Masri, W. S. and Silver, J. R. (1981). Prophylactic anticoagulant therapy in patients with spinal cord injury. *Paraplegia* **19**, 334–42.

El Masri, W., Cochrane, P., and Silver, J. R. (1982). Gastrointestinal bleeding in patients with acute spinal injuries. *Injury* **14**, 162–7.

Epstein, N., Hood, D. D., and Ransohoff, J. (1981). Gastrointestinal bleeding in patients with spinal cord trauma. Effects of steroids, cimetidine, and mini-dose heparin. *J. Neurosurg.* **54**, 16–20.

Faden, A. I., Jacobs, T. P., Patrick, D. H., and Smith, M. T. (1984). Megadose corticosteroid therapy following experimental traumatic spinal injury. *J. Neurosurg.* **60**, 712–17.

Faden, A. I., Jacobs, T. P., Smith, M. T., and Holaday, J. W. (1983). Comparison of thyrotropin-releasing hormone (TRH), naloxone, and dexamethasone treatments in experimental spinal injury. *Neurology* **33**, 673–8.

Foo, D., Subrahmanyan, T. S., and Rossier, A. B. (1981). Post-traumatic acute anterior spinal cord syndrome. *Paraplegia* **19**, 201–5.

Frankel, H. L., Mathias, C. J., and Spalding, J. M. K. (1975). Mechanisms of reflex cardiac arrest in tetraplegic patients. *Lancet* **ii**, 1183–5.

Green, B. A., Kahn, T., and Klose, K. J. (1980). A comparative study of steroid therapy in acute experimental spinal cord injury. *Surg. Neurol.* 13, 91–7.

Gunby, P. (1982). Study shows little effect of steroids in spinal cord injury. *Medical News J. Am. Med. Ass.* **248**, 1035.

Guttmann, L. and Silver, J. R. (1965). Electromyographic studies on reflex activity of the intercostal and abdominal muscles in cervical cord lesions. *Paraplegia* **3**, 1–22.

Hachen, H. J. (1974). Anticoagulant therapy in patients with spinal cord injury. *Paraplegia* **12**, 176–87.

Hall, E. D., Wolf, D. L., and Braughler, J. M. (1984). Effects of a single large dose of methylprednisolone sodium succinate on experimental posttraumatic spinal cord ischemia. Dose–response and time-analysis. *J. Neurosurg.* **61**, 124–30.

Hall, M. (1843). *New memoir on the nervous system.* Baillière, London.

Halloran, L. G., Zfass, A. M., Gayle, W. E., Wheeler, C. B., and Miller, J. D. (1980). Prevention of acute gastrointestinal complications after severe head injury: a controlled trial of cimetidine prophylaxis. *Amer. J. Surg.* **139**, 44–8.

Hansebout, R. R., Tanner, J. A., and Romero-Sierra, C. (1984). Current status of spinal cord cooling in the treatment of acute spinal cord injury. *Spine* **9**, 508–11.

Hardy, A. G. and Rossier, A. B. (1975). *Spinal cord injuries. Orthopedic and neurological aspects.* Thième, Stuttgart.

Kewalramani, L. S. (1979). Neurogenic gastroduodenal ulceration and bleeding associated with spinal cord injuries. *J. Trauma* **19**, 259–65.

Koons, D. D., Gildenberg, P. L., Dohn, D. F., and Henoch, M. (1972). Local hypothermia in the treatment of spinal cord injuries. *Cleveland Clin. Quart.* **39**, 109–17.

Levine, B. A., Schwesinger, W. H., Sirinck, K. R., Jones, D., and Pruitt, B. A. (1978). Cimetidine prevents reduction in gastric mucosal blood flow during shock. *Surgery* **84**, 113–19.

Lewin, M. G., Pappius, H. M., and Hansebout, R. R. (1972). Effects of steroids on edema associated with injury of the spinal cord. In *Steroids and brain edema* (ed. H. J. Reulen and K. Schurmann), pp. 101–12. Springer, New York.

McCarthy, D. M. (1983). Editorial. Ranitidine or cimetidine. *Ann. Int. Med.* **99**, 551–3.

Meacham, W. F. and McPherson, W. F. (1973). Local hypothermia in the treatment of acute injuries of the spinal cord. *South Med. J.* **1**, 95–7.

Meinecke, F. W. (1968). Frequency and distribution of associated injuries in traumatic paraplegia and tetraplegia. *Paraplegia* **5**, 196–209.

Ohry, A., Molho, M., and Rozin, R. (1975). Alterations of pulmonary function in spinal cord injured patients. *Paraplegia* **13**, 101–8.

Osterholm, J. L. (1974). The pathophysiological response to spinal cord injury. *J. Neurosurg.* **40**, 5–33.

Perkash, A., Prakash, V., and Perkash, I. (1978). Experience with the management of thromboembolism in patients with spinal cord injury: Part I. Incidence, diagnosis and role of some risk factors. *Paraplegia* **16**, 322–31.

Priebe, H. J., Skillman, J. J., Bushnell, L. S., Long, P. C., and Silen, W. (1980). Antacid versus cimetidine in preventing acute gastrointestinal bleeding. A randomized trial in 75 critically ill patients. *New Engl. J. Med.* **302**, 426–30.

Rogers, W. A. (1957). Fractures and dislocations of the cervical spine. An end-result study. *J. Bone Joint Surg.* **39**(A), 341–76.

Rossier, A. and Brunner, U. (1964). Zur initialen Behandlung der frischen traumatischen Querschnittsläsion. *Schweiz. med. Wschr.* **94**, 362–70.

Rossier, A. B., Fam, B. A., di Benedetto, M., and Sarkarati, M. (1980). Urethro-vesical function during spinal shock. *Urol. Res.* **8**, 53–65.

Rowed, D. W., McLean, J. A. G., and Tator, C. H. (1976). Somatosensory evoked potentials in acute spinal cord injury: prognostic value. *Surg. Neurol.* **9**, 203–10.

Sandor, F. (1966). Diaphragmatic respiration: a sign of cervical cord lesion in the unconscious patient (horizontal paradox). *Br. Med. J.* **1**, 465–6.

Scher, A. T. (1981). Unrecognized fractures and dislocations of the cervical spine. *Paraplegia* **19**, 25–30.

Schneider, R. C. (1955). The syndrome of acute anterior spinal cord injury. *J. Neurosurg.* **12**, 95–122.

Sherrington, C. S. (1899). On the spinal animal. *Medico-Chirurgical Trans.* **82**, 449–75.

Silver, J. R. (1974). The prophylactic use of anticoagulant therapy in the prevention of pulmonary emboli in one hundred consecutive spinal injury patients. *Paraplegia* **12**, 188–96.

Soderstrom, C. A. and Ducker, T. B. (1985). Increased susceptibility of patients with cervical cord lesions to peptic gastrointestinal complications. *J. Trauma* **25**, 1030–8.

Stauffer, E. S. (1976). Principles of early management of injuries to the spinal cord. In *Current controversies in neurosurgery* (ed. T. P. Morley), pp. 116–22. W. B. Saunders Company, Philadelphia, Pennsylvania.

Tator, C. H. (1976). Treatment of spinal cord injury in the acute stage: the ineffectiveness of cord perfusion. In *Current controversies in neurosurgery* (ed. T. P. Morley), pp. 110–15. W. B. Saunders Company, Philadelphia, Pennsylvania.

Tator, C. H. and Deecke, L. (1972). Normothermic vs hypothermic perfusion in treatment of acute experimental spinal cord injury. *Surg. Forum* **23**, 435–8.

Thienprasit, P., Bantli, H., Bloedel, J. R., and Chou, S. N. (1975). Effect of delayed local cooling on experimental spinal cord injury. *J. Neurosurg.* **42**, 150–4.

Tribe, C. R. (1963). Causes of death in the early and late stages of paraplegia. *Paraplegia* **1**, 19–46.

Wagner, F. C., van Gilder, J. C., and Dohrmann, G. J. (1978). Pathological changes from acute to chronic in experimental spinal cord trauma. *J. Neurosurg.* **48**, 92–8.

Walsh, J. J. and Tribe, C. (1965). Phlebo-thrombosis and pulmonary embolism in paraplegia. *Paraplegia* **3**, 209.

Watson, N. (1968). Venous thrombosis and pulmonary embolism in spinal cord injury. *Paraplegia* **6**, 113–21.

Watson, N. (1978). Anti-coagulant therapy in the prevention of venous thrombosis and pulmonary embolism in the spinal cord injury. *Paraplegia* **16**, 265–9.

Welply, N. C., Mathias, C. J., and Frankel, H. L. (1975). Circulatory reflexes in tetraplegics during artificial ventilation and general anesthesia. *Paraplegia* **13**, 172–82.

White, A. A. and Moss, H. L. (1978). Hangman's fracture with non-union and late cord compression. *J. Bone Joint Surg.* **60**(A), 839–40.

White, R. J. (1986). Historical development of spinal cord cooling. *Surg. Neurol.* **25**, 295–8.

Wolman, L. (1965). The disturbance of circulation in traumatic paraplegia in the acute and late stages: a pathological study. *Paraplegia* **2**, 213–26.

York, D. H., Watts, C., Raffensberger, M., Spagnolia, T., and Joyce, C. (1983). Utilization of somatosensory evoked cortical potential in spinal cord injury. Prognostic limitations. *Spine* **8**, 832–9.

III

Physiological assessment of spinal cord function

8

Assessment of spinal cord function by means of somatosensory-evoked potentials

E. M. Sedgwick

Introduction

In this chapter I review experimental work on spinal cord potentials in animals in so far as it relates to human subjects and the interpretation of somatosensory-evoked potentials in clinical neurophysiology. Cross-species assumptions, usually that man is like the cat, have served us well and led to no fundamental misinterpretations to date, but one must remain aware that new facts may not always fit old assumptions. Our understanding of evoked potentials from the spinal cord derives from experimental work begun in the 1930s, but clinical studies did not begin in earnest until microprocessors became available in the 1970s.

Animal studies

Gasser and Graham (1933) used one of the early oscilloscopes to study potentials recorded from the cord *dorsum* of the cat and provided most of the information about the cord *dorsum* potential. An electrode placed on the *dorsum* of the cord will record two different phenomena: the compound action potential in the dorsal root which travels rostrally in the dorsal *funiculus*, and a cord *dorsum* potential which has a slower onset and longer duration reaching a negative peak in a few milliseconds with a following positive peak of longer duration. This potential is not conducted along the dorsal columns but is confined to the region of the root entry zone (Fig. 8.1).

The cord *dorsum* potential was shown to be uninfluenced by deep ether narcosis, but was abolished by hypoxia before the compound action potential of the dorsal column was lost. Double stimuli to split roots or to adjacent homolateral roots gave a potential which was smaller than the algebraic sum of the roots stimulated singly, indicating that the pathways converged on a common structure. Stimulation of root on the right and left sides, however, gave responses which did sum algebraically, indicating no side-to-side convergence. Double stimuli showed that the negativity following the second stimulus was attenuated for interstimulus intervals of up to 70 msec,

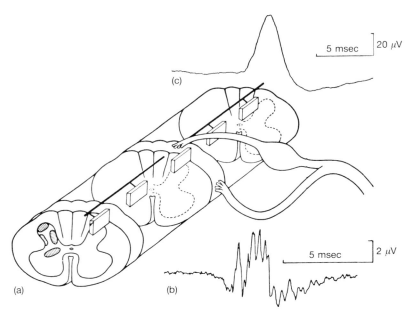

Fig. 8.1. (a) Diagram of the spinal cord with the nerve roots. The heavy line denotes a sensory axon from a Krause corpuscle which branches rostrally and caudally on entering the cord. The branches send collaterals into the dorsal horn where terminal arbors occupy a rectangular packet of space within which they make synaptic contact with dendrites of neurons on laminae III and IV. The origin of the cord *dorsum* potential is from laminae III and IV where the negative sinks are depicted by horizontal hatch in the diagram. The vertical hatch represents laminae I and V within which high threshold (pain) afferents terminate. (b) Recording from an L4 root in man after stimulation of the L4 dermatome. The responses are polyphasic and of long duration. The response latency was about 16 msec; the initial part of the trace is not included. (c) A cord *dorsum* potential in man recorded from an electrode in the epidural space. A single negative–positive wave with a latency of 12 msec is shown. The tibial nerve was stimulated in the popliteal *fossa*. *Note* the different calibrations in (b) and (c). Negativity is denoted by an upward deflection in this and other figures.

although the dorsal column potential recovered from its relative refractory period within 2 msec. Gasser and Graham (1933) went on to show that the cord *dorsum* potential was independent of reflex and antidromic activity in motoneurons. As the potential was produced by neither the first neuron of a reflex arc (primary afferent) nor the last (motoneuron) it was called the intermediary potential and attributed to activity in the synapses, dendrites, and bodies of spinal interneurons.

A number of studies have confirmed and amplified these findings, which were reviewed by Bernhard (1952). In particular it was shown that the cord

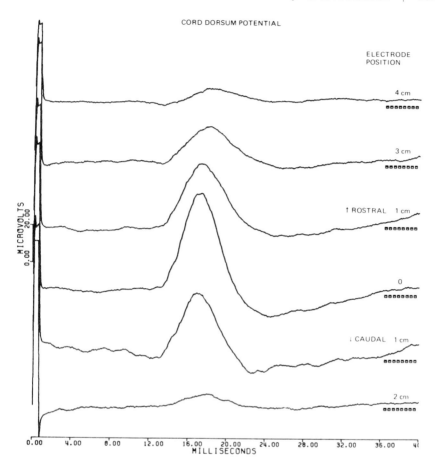

CORD DORSUM POTENTIAL

ELECTRODE
POSITION

4 cm

3 cm

↑ ROSTRAL   1 cm

0

↓ CAUDAL   1 cm

2 cm

Fig. 8.2. Recording from the epidural space of a patient with a T5 spinal injury. The tibial nerve at the popliteal *fossa* was stimulated. The recording electrode was withdrawn for each record and the maximum amplitude was seen when the tip was at the level of the T11 vertebral body, equivalent to the L5/S1 root entry zone of the cord. The potential amplitude falls rostrally and caudally from this point.

dorsum potential was produced by low-threshold cutaneous afferents; low-threshold muscle afferents produced almost no potential. The generator of both the negative and positive phases is localized in the grey matter of the dorsal horn extending up to 30 mm rostral and caudal to the root entry zone (Bernard, Lindblom, and Ottoson 1953). A similar situation obtains in man, as shown in Fig. 8.2 (Dimitrijevic, Holter, Sedgwick, and Sherwood 1986). High stimulus strengths would produce a second and sometimes a third negativity on the trailing edge of the first; this appears to be due to recruitment of higher threshold cutaneous fibres.

Subsequent investigators have used micropipette electrodes to record from the dorsal horns both the field potentials and single-neuron discharges (Beall, Applebaum, Foreman, and Willis 1977). It is apparent that the three negative waves seen from the cord *dorsum* are indeed the result of input from afferent fibres with progressively higher stimulus thresholds, the third negativity being dependent on the A-delta fibres. Figure 8.1 shows the location of the negative sinks within the dorsal horn which identify the sites of the active neurons. The first and second negative potentials have sinks in laminae III and IV of Rexed, while the third, high-threshold, has sinks in the superficial lamina I and another in laminae V–VI.

The effect of spinal cord transection on the cord *dorsum* potential was studied by Lindblom and Ottoson (1953), who reported that the negative phase in decerebrate cats was increased after cord transection by up to 20 per cent but that the rostrocaudal distribution was not changed. Unfortunately, the condition of the animals (blood pressure, etc.) was not monitored in detail, but the interpretation given was of release from a descending tonic inhibition from the brain stem which normally attenuated the cord *dorsum* potential.

The detailed anatomy of the primary afferent terminals and of the neurons in the dorsal horns has been elegantly reviewed by Brown (1981) and shows a pleasing correspondence with the properties of the dorsal horn potential. Many of the terminal arbors of the low-threshold cutaneous primary afferents contact neurons of laminae III and IV where the negativity is at its maximum and whose dendritic fields are dorsoventrally orientated in a fashion similar to the pyramidal neurons in the cerebral cortex. If these cells are the generators of the cord *dorsum* negativity then one may predict a positive potential from the ventral surface of the cord. This is borne out by depth recordings in the cord with microelectrodes (Beall *et al.* 1977), and by recordings in humans with epidural electrodes (Shimoji, Kano, Higashi, Morioka, and Henschel 1972; Beric, Dimitrijevic, Prevec, and Sherwood 1986; Dimitrijevic *et al.* 1986) or with an electrode in the oesophagus which is close to the anterior spinal cord (Desmedt and Cheron 1981). On entering the cord the afferent fibres divide into ascending and descending branches, each of which sends collaterals into the dorsal horn at intervals of about 0.8 mm and together extending in excess of 7 mm in the rostrocaudal direction. The terminal arbors occupy a 'packet' of tissue in laminae III and IV, the shape and orientation of which depends on the receptor type. The terminals of a rapidly adapting receptor, probably a Krause corpuscle, from glabrous skin is illustrated in Fig. 8.1.

Dorsal columns

The simple view of the dorsal columns as a highway of primary afferent fibres ascending to the dorsal column nuclei in the medulla has had to be modified considerably. Only about 25 per cent of the primary afferent fibres from the hindlimb which enter the dorsal column ascend as far as the dorsal column nuclei; a large proportion of the other fibres are post-synaptic from cells in laminae III and IV. Other neurons in these laminae give rise to the spinocervical tract axons which ascend in the dorsolateral *funiculus*; the status of this tract in man is unknown and may be minor (as, in the cat, it carries information concerning hair movement). Also in the dorsolateral *funiculus* are fibres of the dorsal spinocerebellar tract which originate in the cells of Clarke's column and carry information from muscle spindle afferents. This tract is now believed to terminate in nucleus Z, which lies just rostral to the gracile and cuneate nuclei, and to send collaterals to the cerebellum. Nucleus Z is a part of the dorsal column nuclear complex and sends axons into the medial *lemniscus*. Group-I afferents from forelimb muscle spindles ascend in the dorsal columns (York 1985).

The spinothalamic tract seems to make no contribution to the cord *dorsum* potential. Campbell (1985) recorded from the cord surface itself, and even with electrodes deep within the anterolateral quadrant, while stimulating peripheral nerves in patients undergoing anterolateral cordotomy, but found only an ill-defined potential with a latency corresponding to conduction in A-delta fibres.

Lumbar somatosensory-evoked potentials

The cord *dorsum* potential, together with afferent and efferent volleys in the *cauda equina*, may be recorded by skin electrodes (see Figs 8.2 and 8.3). Epidural recordings show these potentials to be up to $50 \mu V$ in amplitude, but surface recordings give very small responses of less than one-tenth that amplitude. They are difficult to record, especially in well-built people and those who cannot relax (Dimitrijevic, Lehmkuhl, Sedgwick, Sherwood, and McKay 1981). The potentials were present and appeared normal after spinal injury some distance above the L5 and S1 segments and were present in spinal shock (El-Negamy and Sedgwick 1978). In some patients abnormalities have been found, sometimes asymmetrically with delay or absence of a component wave even when the cord lesion was well above the lumbosacral region (Lehmkuhl, Dimitrijevic, and Renouf 1984). In a large series with lesions at T8 or above, Beric, Dimitrijevic, and Light (1987) found abnormalities in fifteen out of 130 patients who had no clinical evidence of an additional lesion below T8. On more detailed investigation these patients showed unexpected abnormalities of bladder function, revealed by

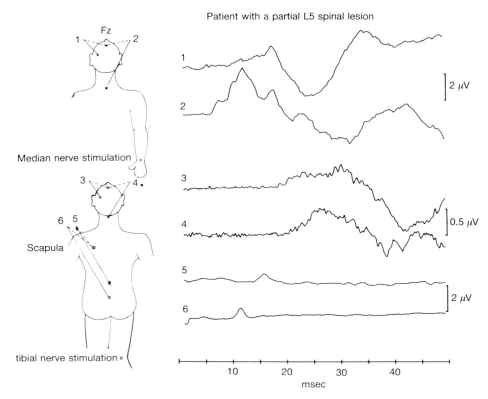

Fig. 8.3. Surface-recorded potentials from a patient with a partial L5 traumatic cord lesion. The upper two traces show normal potentials from the cervical cord and cortex following stimulation of the median nerve. The lowest two traces show a normal *cauda equina* potential (trace 6) depicting the ascending volley from the tibial nerve. Trace 5 shows a cord *dorsum* potential of low amplitude from an electrode on the skin overlying the root entry zone. Traces 3 and 4 show the cortical response following tibial nerve stimulation. It is very low in amplitude, delayed about 10 msec, and spread out in time.

urodynamic studies, such as a flaccid bladder. This is taken as evidence of a second lesion, usually involving the *cauda equina*. The cause is not known, but arachnoiditis has been suggested as has a stretching due to tethering of the cord in the region of the injury followed by contraction of the scar tissue.

The positive wave of the cord *dorsum* potential is thought to be related to pre-synaptic inhibition. Delwaide, Schoenen, and de Pasque (1985) have suggested that lack of pre-synaptic inhibition may be responsible, in part, for spasticity. His findings of a small P-wave in patients with spasticity due to multiple sclerosis appears to support this. However, Dimitrijevic *et al.*

(1986) reported two situations in which pre-synaptic inhibition is enhanced, but there was no accompanying change in the P-wave. A type of spinal inhibition revealed by H-reflex studies and known as D1 (El-Tohamy and Sedgwick 1983) was present after spinal transection in man in most cases (El-Tohamy and Sedgwick 1982).

Monitoring of spinal cord function during surgery

Corrective surgery for scoliosis carries a risk of a 1.17 per cent incidence of neurological complications and a 0.56 per cent risk of partial or complete paraplegia (MacEwen, Bunnell, and Sriram 1975). The reversibility of this neurological morbidity is inversely proportional to the time elapsed before removal of the spinal instrumentation. Monitoring of the spinal cord function can be done intraoperatively by the wake-up test (Vauzelle, Stagnara, and Jouvinroux 1973). However, this is difficult to perform more than once and lengthens the final operative time by approximately fifteen minutes and cannot be applied to deaf and mentally retarded patients. Electrophysiological monitoring has become popular in the last decade and gives a continuous evaluation of the integrity of the dorsal column of the spinal cord. It can be done by recording cortical (Nash, Lorvig, Schatzinger, and Brown 1977; Tsuyama, Tsuzuki, Kurokawa, and Imai 1978) or epidural potentials following stimulation of peripheral nerves in the lower limbs (Engler, Spielholz, Bernhard, Danzinger, Merkin, and Wolff 1978; Jones, Edgar, and Ransford 1982; Maccabee, Levine, Pinkhasov, Cracco, and Tsairis 1983; McWilliam, Connor, and Pollock 1985; see also Grundy 1983). Others prefer a technique in which the stimulating as well as the recording electrodes are placed in the epidural space (Bradshaw, Webb, and Fraser 1984; Whittle, Johnston, Besser, Taylor, and Overton 1984).

The cortical somatosensory-evoked potentials (SSEPs) are affected by anaesthetic agents (particularly halothane), hypotension (Tamaki, Tsuji, Inoue, and Kobayashi 1981), and hypoxia (Clark and Rosner 1973), whereas the epidural responses are relatively stable.

Reviewing the usefulness and reliability of monitoring, Jones, Carter, Edgar, Morley, Ransford, and Webb (1985) concluded that patients with a normal epidural potential throughout surgery did not have major neurological deficit, although up to 1.5 per cent had minor post-operative neurological complications. A high proportion [38 per cent in Jones *et al.* (1985) and up to 100 per cent in other series] of those patients whose potential became abnormal during surgery and had not recovered by the end of surgery had a neurological deficit post-operatively, which amounted to paraplegia or tetraplegia in some cases. In cases where the potential deteriorated and then improved before the end of surgery patients sometimes had minor and transient deficits (see Fig. 8.4). None of these techniques is infallible,

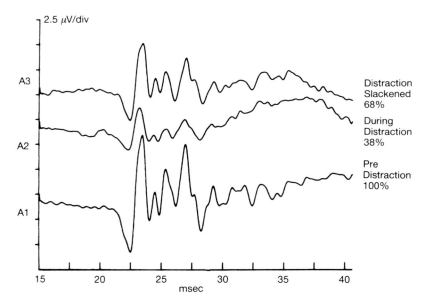

Fig. 8.4. Epidural recording at the T3 level following stimulation of the posterior tibial nerve of the ankle. The 18-year-old patient was undergoing Harrington rod instrumentation with spinal fusion for idiopathic scoliosis. During distraction the amplitude of the ascending volley in the dorsal and dorsolateral *funiculi* decreased (middle trace). The surgeon was informed and slackened the distracting force, after which the potential recovered (upper trace). There was no post-operative neurological deficit.

however, and a recent paper from four centres in the USA reported six patients developing post-operative neurological deficit without any change in the intraoperatively recorded cortical-evoked responses after spinal surgery for scoliosis (one case) or other conditions (five cases). The authors comment that they appear to be false negatives for medical rather than for technical reasons, as widely different stimulus and recording parameters were used at the four centres (Nordwall, Axelgaard, Harada, Valencia, McNeal, and Brown 1979). Conduction in the posterior column and dorsal spinocerebellar pathways is monitored by recording epidural responses, so damage to the more ventrally placed motor pathways may go undetected. Percutaneous stimulation of the motor cortex by electrical or magnetic means and recording from the spinal cord has been performed recently to test the integrity of the motor tracts (Boyd, Cowan, Rothwell, Webb, and Marsden 1985).

Our own experience has been with surgery for scoliosis, and we have found that damage is more likely to occur on the side of the concavity, but advise that both legs be stimulated alternately as unilateral changes in the potential may occur. When monitoring damaged cords the potential may be very small or absent above the lesion. In these cases it has been suggested that a fall in potential of only 10 per cent may be significant in this situation, as in one series (Tamaki *et al.* 1981) two patients with a 10 per cent fall in amplitude were neurologically worse post-operatively. In the same series a rise in potential followed removal of spinal compression, usually caused by a tumour, and heralded neurological improvement in nearly all cases. Failure of the potential to change did not preclude neurological improvement, which was observed in 33 per cent of these cases. Loss of the scalp-recorded SSEP during surgery due to an associated spinal cord lesion, a small dorsal horn infarct, has been demonstrated by Shukla, Docherty, Jackson, and Sedgwick (1987).

Scalp-recorded somatosensory-evoked potentials

Scalp electrodes over the appropriate sensory cortex will record a response to stimulation which is variable between subjects but constant for any one subject. Classically it consists of a W-shaped wave with an initial negative peak when recorded after stimulation of the forelimb. The SSEP is produced by stimuli travelling in the dorsal *funiculi* and, if a mixed nerve is stimulated, by the muscle spindle afferents, rather than by the slower conducting cutaneous fibres (Burke, Skuse, and Lethlean 1981). Normal SSEPs are produced by direct stimulation of the dorsal column (Larson, Sances, and Christenson 1966), and clinical studies show that patients with pain and temperature deficits due to spinal lesions have normal cortical SSEPs whereas those with any proprioceptive loss show an abnormality (Halliday and Wakefield 1963; Giblin 1964). The cortical SSEP can therefore be recorded and used as a measure of dorsal column function.

The scalp-recorded SSEP has been under close investigation as a means of assessing the extent and prognosis of traumatic spinal lesions. It is agreed by all that patients with a clinically complete lesion have no SSEP recordable after simulation below the lesion. In fact, if an SSEP were found the lesion would have to be redefined as incomplete (Fig. 8.3).

Recording in the acutely injured reveals all possible alterations of the SSEP. It may be absent initially, but then reappear, sometimes recovering to normal within about 2–3 weeks. Sometimes a recovering potential will deteriorate temporarily about five days after injury. Occasionally an SSEP will deteriorate during the first days after injury and never reappear, but a slow and fluctuating recovery may be seen (Rowed, MacLean, and Tator 1978; Spielholz, Benjamin, Engler, and Ranschoff 1979; Perot and Vera 1982; Stohr, Buettner, Riffel, and Koletyki 1982; Young 1982; York, Watts, Raffensberger, Spagnola, and Joyce 1983; Schiff, Cracco, Rossini, and Cracco 1984; Chabot, York, Watts, and Waugh 1985; Keim, Hajda,

Gonzales, Brand, and Balasubramanian 1985; Pierlovisi-Lavaire, de Buschop, Brant, Chapuis-Ducoffre, Bence, and Millett 1985; Young 1985).

The SSEP correlates with sensory function but, after injury, a returning SSEP may precede recovery of sensory function. One never finds preserved sensory function with an absent SSEP in the early weeks, although this situation may occur late in recovery and is not infrequently observed in patients with chronic cord lesions due to multiple sclerosis. The converse of a preserved or recovered SSEP without sensory function is frequent soon after injury, but should be followed by sensory recovery. Once the chronic stage is reached, after 2–3 months, a recovered SSEP without sensation suggests that there may be a hysterical or malingering overtone to the situation (Sedgwick 1986).

The components of the SSEP which are of most prognostic significance with respect to sensation are those occurring during the first 80 msec after stimulation, and a high-amplitude initial positive wave is a good sign for recovery. The more complex the wave-form and the higher the amplitude, the more complete sensory recovery is likely to be eventually. There was found to be a good correlation between the SSEP on admission and sensory function assessed at discharge ($r = 0.64$). The correlation with motor function at discharge was also found to be positive but less secure at 0.44 (Ziganow 1986). In some cases the SSEP consists of a late (>100 msec) broad negative wave which correlates poorly with sensory function and tends to disappear if the short-latency SSEP components begin to return. In practice it is difficult to be sure when a potential is absent.

In a series of chronic patients recorded seven months to twenty-eight years after injury, Dimitrijevic, Prevec, and Sherwood (1983) attempted to correlate specific features of sensory testing with the degree of abnormality of the SSEP. Correlation with touch, proprioception, pain, or two-point discrimination was imperfect. It is possible in chronic cases to find some recovery of sensory function but for the SSEP to remain absent. In such cases the SSEP may be more diligently sought, by, for instance, stimulation of the sacral dermatomes or by stimulating at a much slower rate. Dimitrijevic et al. (1981) have shown how the amplitude of the SSEP may increase if very slow rates of stimulation are used.

An abnormal SSEP without clinical sensory change is a frequent finding in patients with cord lesions of other pathologies. In multiple sclerosis, for instance, an abnormal SSEP may be evidence of a paraclinical lesion or may confirm indefinite subjective sensory symptoms (Halliday 1982). Compressive lesions usually produce early abnormalities, whereas intramedullary lesions, even those infiltrating the dorsal columns, may leave the SSEP normal (Maugiere, Ibanez, and Fischer 1985).

Several groups have shown that stimulation of nerves or of the skin at more and more rostral sites eventually gives a normal SSEP, thus identifying the level of the lesion. I have never been able to envisage circumstances in which this would be the only or the best means of determining the level of the lesion. The procedure may have a role in identifying incomplete, diffuse, or multiple lesions, such as occur in multiple sclerosis (Jorg, Dullberg, and Koeppen 1982).

Dermatomal somatosensory-evoked potentials

Cauda equina lesions and root compression are more effectively studied by use of the dermatomal somatosensory-evoked potential (DSSEP). Sedgwick, Katifi, Docherty, and Nicpon (1985) and Katifi and Sedgwick (1986, 1987) have shown that stimulation of the skin at the signature area of a dermatome produces a scalp-recorded SSEP at a latency which is determined by the subject's height. Abnormal, that is, delayed or absent, SSEPs can be recorded in patients with monoradiculopathies: the overlapping innervation of dermatomes seems not to compensate for damage to a single root in the lumbosacral region (see Fig. 8.5). In a series of twenty patients with surgically proven root compressions the DSSEP result indicated the correct diagnosis in nineteen of them. It should be possible to make the DSSEP the primary investigation, replacing myelography, for those patients who will be managed conservatively (Scarff, Dallmann, and

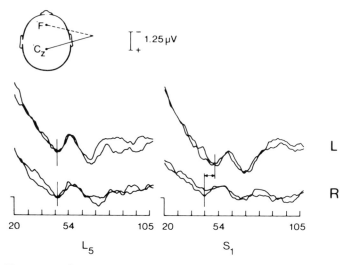

Fig. 8.5. Dermatomal somatosensory-evoked potential recorded from the scalp of a 28-year-old man with back pain and sciatica. Myelography was normal, but the potential from the left S1 dermatome was delayed. At surgery the left S1 root was bound with adhesions and fibrosis.

Bunch 1981; Aminoff, Goodin, Parry, Barbaro, Weinstein, and Rosenblum 1985; Dvonch, Scarff, Bunch, Smith, Lebarge, and Ibrahim 1985; Katifi and Sedgwick 1987).

Brachial plexus

Patients with brachial plexus injuries may have their nerve roots avulsed from the cord, in which case there will be no recovery of function, or may have more distal lesions, in which case some recovery of shoulder movement

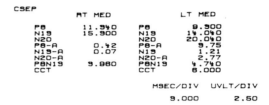

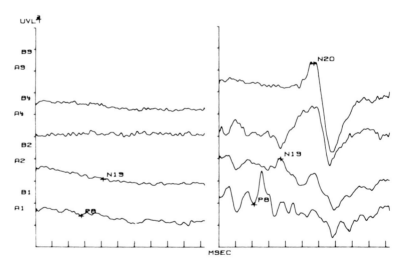

Fig. 8.6. Somatosensory-evoked potentials taken from a patient with a brachial plexus injury. The median nerve at the wrist was stimulated and no responses were seen from the right, but normal responses were recorded from the left. The nerves on the right had undergone Wallerian degeneration, indicating that the lesion was distal to the dorsal root ganglion. Bottom trace (1): recording from Erb's point; trace (2): from over the cervical cord showing the cord *dorsum* potential; upper trace (4): from the somatosensory area of the opposite cortex.

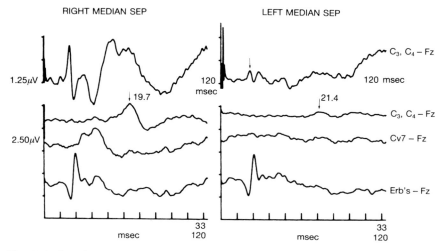

RIGHT MEDIAN SEP   LEFT MEDIAN SEP

Fig. 8.7. Another patient with a brachial plexus injury showing normal responses from the right median nerve. From the left, conduction is normal up to Erb's point, but no cord *dorsum* potential can be seen and the presence of a very small, delayed cortical potential is shown. The findings confirm root avulsion; the lesion was proximal to the dorsal root ganglion. The T1 root was intact clinically and probably accounts for the small cortical potential due to muscle afferents in the median nerve from the abductor *pollicis brevis* muscle, which was functioning.

may be anticipated. Full assessment of a brachial plexus injury involves recording denervation potentials in muscles of the hand, arm, shoulder, and paraspinal region. If the roots have been avulsed the paraspinal muscles and *serratus* anterior will show denervation. The sensory nerves, however, will function perfectly, as they remain attached to their cell bodies in the dorsal root ganglia. Sensory action potentials can be recorded along the nerve as far as Erb's point, but no cord *dorsum* potential and no cortical SSEP is present (Jones 1979) (Figs 8.6 and 8.7). These studies can be performed two weeks after injury. It takes six days for Wallerian degeneration to stop conduction in severed nerves, and somewhat longer for fibrillation potentials to develop in denervated muscles, especially the more distal ones. The main use of these studies is prognostic, but they can be helpful if any form of nerve surgery or grafting is contemplated. Again, it is possible to avoid myelography, which is often used to seek a myelocoele at the point of root avulsion.

Conclusion

Neurological deficits must be understood and described in physiological terms before management and rehabilitation can be applied with logic. Evoked potential techniques offer a new dimension to physiological study of the spinal cord and somatosensory pathways in man. An afferent volley may be followed rostrally into the *cauda equina*, to the first synapse in the dorsal horn, along the ascending dorsal and dorsolateral *funiculi*, through the subcortical structures, and into the sensory cortex where sensory processing begins, and, by waiting a fraction of a second longer, electrical events accompanying perception and cognition may be observed. The techniques may be used to enhance the precision of assessment of a lesion and to follow the evolution of post-traumatic pathophysiology or of any restorative treatment. Most important, they provide a means of testing a number of hypotheses concerning the pathophysiology of neurological deficit. How strange that so few spinal injury centres have seized the opportunities offered.

Acknowledgements

The help and collaboration of many colleagues has been greatly appreciated: Dr J. D. Cole, Dr M. R. Dimitrijevic, Mr T. B. Docherty, Dr H. L. Frankel, Dr L. S. Illis, Dr H. A. Katifi, Dr E. El-Negamy, and Dr A. El-Tohamy.

References

Aminoff, M. J., Goodin, D. S., Parry, G. J., Barbaro, N. M., Weinstein, P. R., and Rosenblum, M. D. (1985). Electrophysiologic evaluation of lumbosacral radiculopathies: electromyography, late responses, and somatosensory evoked potentials. *Neurology (Minneapolis)* **35**, 1514–18.

Beall, J. E., Applebaum, A. E., Foreman, R. D., and Willis, W. D. (1977). Spinal cord potentials evoked by cutaneous afferents in the monkey. *J. Neurophysiol.* **40**, 199–211.

Beric, A., Dimitrijevic, M. R., and Light, J. K. (1987). A clinical syndrome of rostral and caudal spinal cord injury: neurological, neurophysiological and urodynamic evidence for occult sacral lesion. *J. Neurol. Neurosurg. Psychiat.* **50**, 600–6.

Beric, A., Dimitrijevic, M. R., Prevec, T. S., and Sherwood, A. M. (1986). Epidurally recorded somatosensory evoked potential in humans. *Electroencephal. Clin. Neurophysiol.* **65**, 94–101.

Bernhard, C. G. (1952). The cord dorsum potentials in relation to peripheral source of afferent stimulation. *Cold Spring Harbor Symp. Quant. Biol.* **17**, 221–32.

Bernhard, C. G., Lindblom, U. F., and Ottoson, J. O. (1953). The longitudinal distribution of the negative cord dorsum potential following stimulation of low threshold cutaneous fibres. *Acta Physiol. Scand.* **29** (Suppl. 106), 170–9.

Boyd, S. G., Cowan, J. M. A., Rothwell, J. C., Webb, P. J., and Marsden, C. D. (1985). Monitoring spinal motor tract function using cortical stimulation: a preliminary report. In *Spinal cord monitoring* (ed. J. Schramm and S. J. Jones), pp. 227–30. Springer, Berlin.

Bradshaw, K., Webb, J. K., and Fraser, A. M. (1984). Clinical evaluation of the spinal cord monitoring in scoliosis surgery. *Spine* **9**, 636–43.

Brown, A. G. (1981). *Organisation in the spinal cord.* Springer, Berlin.

Burke, D., Skuse, N. S., and Lethlean, A. K. (1981). Cutaneous and muscle afferent components of the cerebral potential evoked by electrical stimulation of human peripheral nerves. *Electroencephal. Clin. Neurophysiol.* **51**, 579–88.

Campbell, J. A. (1985). Observations on somatosensory evoked potentials recorded from within the human spinal cord. PhD thesis, University of Liverpool, Liverpool.

Chabot, R., York, D. H., Watts, C., and Waugh, W. A. (1985). Somatosensory evoked potentials in normal subjects and spinal cord injured patients. *J. Neurosurg.* **63**, 544–51.

Clark, D. L. and Rosner, B. S. (1973). Neurophysiological effects of general anaesthetics. I. The electroencephalogram and sensory evoked responses in man. *Anesthesiology* **38**, 564–82.

Delwaide, P. J., Schoenen, J., and de Pasque, V. (1985). Lumbosacral spinal evoked potentials in patients with multiple sclerosis. *Neurology* **35**, 174–9.

Desmedt, J. E. and Cheron, G. (1981). Prevertebral (oesophageal) recording of subcortical somatosensory evoked potentials in man: the spinal P13 component and the dual nature of the spinal generators. *Electroencephal. Clin. Neurophysiol.* **52**, 257–75.

Dimitrijevic, M. R., Prevec, T. S., and Sherwood, A. M. (1983). Somatosensory perception and cortical evoked potentials in established paraplegia. *J. Neurol. Sci.* **60**, 253–65.

Dimitrijevic, M. R., Holter, J., Sedgwick, E. M., and Sherwood, A. M. (1986). Spatial distribution of the cord dorsum potential and its presence during H reflex inhibition in man. *J. Physiol.* **382**, 72.

Dimitrijevic, M. R., Lehmkuhl, L. D., Sedgwick, E. M., Sherwood, A. M., and McKay, W. B. (1981). Characteristics of spinal cord evoked response in man. *Appl. Neurophysiol.* **43**, 118–27.

Dvonch, V., Scarff, T., Bunch, W. H., Smith, D., Lebarge, H., and Ibrahim, I. (1985). Dermatomal somatosensory evoked potentials: their use in lumbar radiculopathy. *Spine* **9**, 291–3.

El-Negamy, E. and Sedgwick, E. M. (1978). Properties of a spinal somatosensory evoked potential recorded in man. *J. Neurol. Neurosurg. Psychiat.* **41**, 762–8.

El-Tohamy, A. and Sedgwick, E. M. (1982). Spinal inhibitory mechanisms in spasticity. *Electroencephal. Clin. Neurophysiol.* **53**, 1–2P.

El-Tohamy, A. and Sedgwick, E. M. (1983). Spinal inhibition in man: depression of the soleus H reflex by stimulation of the nerve to the antagonist muscle. *J. Physiol.* **336**, 497–508.

Engler, G. L., Spielholz, H. J., Bernhard, W. N., Danziger, F., Merkin, H., and Wolff, T. (1978). Somatosensory evoked potentials during Harrington instrumentation for scoliosis. *J. Bone Joint Surg.* **60**(B), 528–32.

Gasser, H. R. and Graham, H. T. (1933). Potentials produced in the spinal cord by stimulation of dorsal roots. *Amer. J. Physiol.* **103**, 303–20.

Giblin, D. R. (1964). Somatosensory evoked potentials in healthy subjects and in patient with lesions of the nervous system. *Ann. NY Acad. Sci.* **112**, 93–142.

Grundy, B. L. (1983). Intraoperative monitoring of sensory-evoked potentials. *Anesthesiology* **58**, 72–87.

Halliday, A. M. (1982. *Evoked potentials in clinical testing.* Churchill Livingstone, Edinburgh.

Halliday, A. M. and Wakefield, G. S. (1963). Cerebral evoked potentials in patients with associated sensory loss. *J. Neurol. Neurosurg. Psychiat.* **26**, 211–19.

Jones, S. J. (1979). Investigation of brachial plexus traction lesion by peripheral and spinal somatosensory evoked potentials. *J. Neurol. Neurosurg. Psychiat.* **42**, 107–16.

Jones, S. J., Edgar, M. A., and Ransford, A. O. (1982). Sensory nerve conduction in the human spinal cord: epidural recordings made during surgery. *J. Neurol. Neurosurg. Psychiat.* **45**, 446–51.

Jones, S. J., Carter, L., Edgar, M. A., Morley, T., Ransford, A. O., and Webb, P. J. (1985). Experience of spinal cord monitoring in 410 cases. In *Spinal cord monitoring* (ed. J. Schramm and S. J. Jones), pp. 215–20. Springer, Berlin.

Jorg, J., Dullberg, W., and Koeppen, S. (1982). Diagnostic value of segmental somatosensory evoked potentials in cases with chronic progressive para- or tetraspastic syndromes. In *Clinical applications of evoked potentials in neurology* (ed. J. Courjon, F. Maugiere, and M. Revol), pp. 34–58. Raven Press, New York.

Katifi, H. A. and Sedgwick, E. M. (1986). Somatosensory evoked potentials from posterior tibial nerve and lumbo-sacral dermatomes. *Electroencephal. Clin. Neurophysiol.* **65**, 249–59.

Katifi, H. A. and Sedgwick, E. M. (1987). Evaluation of the dermatomal somatosensory evoked potential in the diagnosis of lumbo-sacral root compression. *J. Neurol. Neurosurg. Psychiat.* **50**, 1211–15.

Keim, H. A., Hajda, M., Gonzales, E. G., Brand, L., and Balasubramanian, E. (1985). Somatosensory evoked potentials as an aid in the diagnosis and intraoperative management of spinal stenosis. *Spine* **10**, 338–44.

Larson, S. J., Sances, A., and Christenson, P. C. (1966). Evoked somatosensory potentials in man. *Arch. Neurol.* **15**, 88–93.

Lehmkuhl, D., Dimitrijevic, M. R., and Renouf, F. (1984). Electrophysiological characteristics of lumbosacral evoked potentials in patients with established spinal cord injury. *Electroencephal. Clin. Neurophysiol.* **59**, 142–55.

Lindblom, U. F. and Ottoson, J. O. (1953). Localization of the structure generating the negative cord dorsum potential evoked by stimulation of low threshold cutaneous fibres. *Acta Physiol. Scand.* **106** (Suppl. 29), 191–208.

Maccabee, P. J., Levine, D. B., Pinkhasov, E. I., Cracco, R. Q., and Tsairis, P. (1983). Evoked potentials recorded from the scalp and spinous processes during spinal column surgery. *Electroencephal. Clin. Neurophysiol.* **56**, 569–82.

MacEwen, G. D., Bunnell, W. P., and Sriram, K. (1975). Acute neurological complications in the treatment of scoliosis: a report of the scoliosis research society. *J. Bone Joint Surg.* **57**(A), 404–8.

Mauguiere, F., Ibanez, V., and Fischer, G. (1985). Les potentiels évoques somesthésiques dans les tumeurs intra-rachidiennes. *Rev. Electroencephal. Neurophysiol.* **15**, 95–106.

McWilliam, R. C., Connor, A. N., and Pollock, J. C. S. (1985). Cortical somatosensory evoked potentials during surgery for scoliosis and coarctation of aorta. In *Spinal cord monitoring* (ed. J. Schramm and S. J. Jones), pp. 167–72. Springer, Berlin.

Nash, C. L. Jr., Lorvig, R. A., Schatzinger, L. A., and Brown, R. H. (1977). Spinal cord monitoring during operative treatment of the spine. *Clin. Orthoped.* **126**, 100–5.

Nordwall, A., Axelgaard, J., Harada, Y., Valencia, P., McNeal, D. R., and Brown, J. C. (1979). Spinal cord monitoring using evoked potentials recorded from feline vertebral bone. *Spine* **4**, 486–94.

Perot, P. L. and Vera, C. L. (1982). Scalp-recorded somatosensory evoked potentials to stimulation of nerves in the lower extremities and evaluation of patients with spinal cord trauma. *Ann. NY Acad. Sci.* **388**, 359–68.

Pierlovisi-Lavaire, M., de Buschop, G., Brant, A., Chapuis-Ducoffre, M. F., Bence, Y., and Millet, Y. (1985). Valeur pronostique des potentiels évoques somesthésiques précose au cours des lésions médullaires traumatiques. *Rev. Electroencephal. Neurophysiol.* **15**, 77–83.

Rowed, D. W., McLean, J. A. G., and Tator, C. H. (1978). Somatosensory evoked potentials in acute spinal injury: prognostic value. *Surg. Neurol.* **9**, 203–9.

Scarff, T. B., Dallmann, D. E., and Bunch, W. H. (1981). Dermatomal somatosensory evoked potentials in the diagnosis of lumbar root entrapment. *Surg. Forum* **32**, 489–91.

Schiff, J. A., Cracco, R. O., Rossini, P. M., and Cracco, J. B. (1981). Spine and scalp somatosensory evoked potentials in normal subjects and patients with spinal cord disease: evaluation of afferent transmission. *Electroencephal. Clin. Neurophysiol.* **59**, 374–87.

Sedgwick, E. M. (1986). Simulated paraplegia: an occasional problem for the neurosurgeon. *J. Neurol. Neurosurg. Psychiat.* **49**, 336.

Sedgwick, E. M., Katifi, H. A., Docherty, T. B., and Nicpon, K. (1985). Dermatomal somatosensory evoked potentials in lumbar disc disease. In *Evoked potentials. Neurophysiological and clinical aspects* (ed. C. Morocutti and P. A. Rizzo), pp. 77–88. Elsevier, Amsterdam.

Shimoji, K., Kano, T., Higashi, H., Morioka, T., and Henschel, E. O. (1972). Evoked spinal electrograms recorded from epidural space in man. *J. Appl. Physiol.* **33**, 468–71.

Shukla, R., Docherty, T. B., Jackson, R. K., and Sedgwick, E. M. (1987). Comparison of epidural responses from right and left posterior tibial nerves during surgery for scoliosis. *Electroencephal. Clin. Neurophysiol.* **67**, 6P.

Spielholz, N. I., Benjamin, M. V., Engler, G., and Ranschoff, J. (1979). Somatosensory evoked potentials and clinical outcome in spinal cord injury. In *Neural trauma* (ed. A. J. Popp, R. S. Bourke, L. R. Nelson, and H. K. Kimelberg), pp. 217–27. Raven Press, New York.

Stohr, M., Buettner, V. W., Riffel, B., and Koletyki, E. (1982). Spinal somatosensory evoked potentials in cervical cord lesions. *Electroencephal. Clin. Neurophysiol.* **54**, 257–65.

Tamaki, T., Tsuji, H., Inoue, S. I., and Kobayashi, H. (1981). The prevention of iatrogenic spinal cord injury utilising the evoked spinal cord potential. *Internat. Orthoped.* **4**, 313–17.

Tsuyama, N., Tsuzuki, N., Kurokawa, T., and Imai, T. (1978). Clinical application of spinal cord action potential measurement. *Internat. Orthoped.* **2**, 39–46.

Vauzelle, C., Stagnara, P., and Jouvinroux, P. (1973). Functional monitoring of the spinal cord during spinal surgery. *Clin. Orthoped.* **93**, 173–8.

Whittle, I. R., Johnston, I. H., Besser, M., Taylor, T. K. F., and Overton, J. (1984). Intraoperative spinal cord monitoring during surgery for scoliosis using somatosensory evoked potentials. *Aust. NZ J. Surg.* **54**, 553–7.

York, D. H. (1985). Somatosensory evoked potentials in man: differentiation of spinal pathways responsible for conduction from the forelimb vs hindlimb. *Progr. Neurobiol.* **25**, 1–25.

York, D. H., Watts, C., Raffensberger, M., Spagnola, T., and Joyce, C. (1983). Utilization of somatosensory evoked cortical potentials in spinal cord injury. *Spine* **8**, 832–9.

Young, W. (1982). Correlation of somatosensory evoked potentials and neurological findings in spinal cord injury. In *Early management of acute spinal cord injury* (ed. C. H. Tator), pp. 153–65. Raven Press, New York.

Young, W. (1985). Cortical somatosensory evoked potential changes in spinal cord injury. In *Proceedings of the Third International Symposium on Spinal Cord Monitoring* (ed. J. Schramm), pp. 121–42. Springer, Berlin.

Ziganow, S. (1986). Neurometric evaluation of the cortical somatosensory evoked potential in acute incomplete spinal cord injuries. *Electroencephal. Clin. Neurophysiol.* **65**, 86–93.

9

The pathophysiology of the autonomic nervous system in spinal cord injury

J. D. Cole

Progress in understanding and in assessment in internal medicine has often occurred at two levels. At the first, clinical practice has sought to obtain information with minimal invasiveness and reasonable speed. Clinical research, on the other hand, has been more detailed and has necessarily involved more prolonged investigation. This division has been especially evident in the clinical syndromes of autonomic failure, and particularly so in spinal cord injury for conceptual and methodological reasons.

This review will be concerned with the results of research in the field of autonomic pathophysiology after spinal cord injury, although clinical assessment techniques will also be considered. After a short historical introduction the *sequelae* of spinal injury will be discussed, beginning with the circulatory system. The urogenital system will not be considered since it is the subject of Chapter 10. Finally, a short consideration of the physiological mechanisms underlying phenomena like autonomic dysreflexiae will be given.

Historical introduction

Injury to the spinal cord has usually been considered in terms of the resultant loss of movement—tetra- and paraplegia. Whilst sensory loss has also been considered the third aspect of deranged function, autonomic failure, has been less well documented. There was, however, a mention of lost voluntary control of micturition in the Edwin Smith papyrus (Egyptian *c*.5000 BC; see Verkuyl 1976). With a longer survival time post-injury the loss of control of circulatory reflexes in tetraplegia, as well as the less well understood functions concerned, say, with the splanchnic organs, can have important consequences in rehabilitation and the quality of life. Hilton (1860) was the first to describe the signs of autonomic dysreflexiae (see Frankel and Mathias 1979), while Bowlby (1890) reported a C7 tetraplegic who twenty weeks after his injury had a tingling sensation over his chest with profuse perspiration of the head and neck on passage of a urinary catheter, and

Hutchinson (1875) had noted the effect on temperature regulation of spinal injury.

The next advances in thinking on spinal injury awaited the observations of several physician–physiologists on the large numbers of men injured during the First World War. Holmes (1915) served as a field officer and noted how recently injured tetraplegics were at risk from hypothermia. They had not only lost effective temperature regulation but also many of the signs of hypothermia as well. However, the major papers of this period were based on observations of patients a little later after their injury and were made at the Empire Hospital, London, by Head and Riddoch (1917) and Riddoch (1917).

They comment in their introductions that their interests merged at the bladder, Riddoch being interested in the phenomenology of spinal cord injury while Head's main concern was to understand the pathophysiology of the excessive sweating found in tetraplegics recovered from the initial phase of spinal shock.

They found that the automatic bladder in such patients could be made to evacuate its contents at lower volumes than was usual by provoking a flexor spasm, by a single deep breath or by stroking the glans penis. Evacuation of the bladder could in turn lead to a flexor spasm. Their conclusion from this was that afferents from below the lesion had profound effects on reflexes not normally within their influence. The stimuli for excessive sweating were urinary retention, evacuation or catheterization, a rectal enema, or, once more, flexor spasm, all stimuli below the lesion. The area of the body affected by the sweating, which they distinguished from the generalized sweating associated with pyrexia, depended on the level of the lesion. Thus in cervical lesions it included the head, neck, and upper limbs, whilst in T3 lesions the arms and part of the trunk were affected, but not the head. (Not all excessive sweating involved fibres from exclusively below the lesion: in a complete T2 lesion sweating to bladder distension involved the head.) They compared this distribution with that of the sympathetic outflow which had been described by Langley in 1894. They concluded that the pathological hyperhidrosis was brought about through the sympathetic system. In addition to the cause of sweating being afferent activity below the lesion, so was the sympathetic outflow caudal to the spinal damage. They discussed the affects of loss of descending control as being in terms of a loss of 'local sign' in reflex activity. Its consequences were that a small afferent input led to far more widespread effects than would normally occur. This they termed the 'mass reflex'. They made no mention of the effects of autonomic dysreflexiae on pulse or blood pressure, the use of sphygmomanometers not being routine until several decades later (Whitteridge, pers. comm.). Riddoch (1917) also noted that during spinal shock the skin became pale and cold, and was more prone to bed sores. With the return of reflex activity the colour improved and sores sometimes had a chance to heal. The implication was that during the arreflexic stage of spinal shock there was reduced effective vasomotor activity, something confirmed subsequently by Mathias, Christensen, Frankel, and Spalding (1979a).

The involvement of the cardiovascular system in autonomic dysreflexiae was not firmly established until the classical studies of Guttman and Whitteridge (1947) at Stoke Mandeville on patients with complete lesions from the Second World War. They described the sequence of clinical symptoms following controlled urinary bladder distension, including, initially, facial flushing and sweating, tightness in the chest, throbbing in the throat, blockage of the nasal passages, and pupillary dilatation. This sweating tended to be absent in patients with mid-thoracic lesions and below. There was a differential fall in temperature of the body below the lesion and a rise above, compatible with a defect in thermoregulation. However, it also suggested to Guttman and Whitteridge that a redistribution of blood was occurring and alerted them to the need to look at the cardiovascular effects of their manœuvre.

In all patients with T5 lesions and above, bladder distension was associated with severe rises in both systolic and diastolic blood pressure and with bradycardia. In those with T9–10 lesions, slight rises were seen, and in patients with lesions below T10 none. The cut-off point was at T5. It is interesting to note in their Fig. 17 (Guttman and Whitteridge 1947) that hypertension was greatest in those patients with high thoracic lesions, suggesting that some compensation was possible if the thoracic cord (and sympathetic outflow) was intact (see the section on postural hypotension below).

Vasoconstriction occurred in the fingers of those patients who had severe hypertension, via sympathetic fibres arising from below the lesion, although other experiments (Gilliatt, Guttman, and Whitteridge 1947) had shown that cutaneous vasomotor reflexes could be elicited in some low cervical lesions from stimuli above the lesion. This suggested that blood vessels in the fingers may be under neural control from several segments of the cord. For this reason Guttman and Whitteridge (1947) felt unable to say whether the hypertension was due to lack of compensatory vasomotor control to the splanchnic bed alone or whether that to the upper limb was also of importance.

Guttman and Whitteridge (1947) concluded their paper by suggesting that the autonomic reflex responses observed represented abnormal activity of a viscus below the lesion and that their correct recognition should lead to immediate and appropriate action. The modern era of logical treatment and understanding of autonomic dysreflexia had begun.

Cardiovascular *sequelae*

During acute spinal shock

Soon after spinal cord injury the resting blood pressure in tetraplegics is lower than in normal subjects (Walsh 1960; Meinecke, Rosenkrantz, and Kurek 1971). As Mathias and Frankel (1983*a*) point out, measurements of blood pressure immediately after injury have not been made for obvious reasons. There is some evidence from animal work that there is a short period of hypertension after injury, possibly because of a brief period of excess catecholamine secretion. Meinecke *et al.* (1971) studied blood pressure in a group of patients from post-injury day one. Initially all patients when horizontal had low blood pressure. Those with L1 lesions and below recovered to normal by day four. Those with low thoracic and lumbar injuries were scarcely affected subsequently. Those with high thoracic and cervical lesions showed postural hypotension (see below). The basal heart rate was found to be below 100 beats/min in tetraplegics, but higher in those with lower lesions. The difference may reflect reduced hormonal and neural sympathetic activity in the former group.

In addition to hypotension there is also evidence that during arreflexic spinal shock there is little significant cardiovascular reflex activity. Mathias *et al.* (1979*a*) studied five recent (less than two weeks post-injury) C5 tetraplegics whilst in shock and compared them with chronic C5 tetraplegics and a series of normal controls. The recently injured had a lower mean blood pressure and particularly a lower diastolic than controls, with little subsequent increase in the chronic group. Bladder stimulation in chronic tetraplegics led to hypertension, bradycardia, and a rise in plasma noradrenalin, but not adrenalin (which is against a co-ordinated involvement of the adrenal medulla in this response). In the acutely injured patients bladder stimulation only led to the most minimal changes. In an earlier study, Mathias (1976) had shown that both adrenalin and noradrenalin were low in recently injured tetraplegics at rest. This suggests that little significant cardiovascular reflex activity occurs during spinal shock, although Silver (1970) showed that some cutaneous reflexes may be elicitable during this stage.

After the initial phase of spinal shock: postural hypotension

Recently injured tetraplegics with lesions at T5 and above have a profound hypotension if raised from a horizontal position. In his review of this, Guttman (1976) reported that, with head-up tilt, patients reported such symptoms as visual blurring, and then loss of vision, giddiness, tinnitus (interestingly, all vertebrobasilar symptoms), and parathesiae as accompaniments of cerebral hypoxia secondary to hypotension. Syncope also occurred, not infrequently.

Rowell (1984) has suggested that on assuming an upright posture 70 per cent of the blood volume is below the heart, and 75 per cent of this is in the venous capacitance system. In these circumstances total peripheral resistance must be raised if blood pressure is to be maintained. Normally this is achieved by increased vasoconstriction, both in the splanchnic area and elsewhere in a co-ordinated reflex response. Its afferent loop involves cardiopulmonary baroreceptors responding to falls in central venous pressure. However, even in intact man, blood pressure can fall after a period of upright posture if muscular contraction is not available to assist venous return. After spinal cord injury postural hypotension has been attributed to a loss of co-ordinated sympathetic activity to raise pressure. This effect is seen most in T5 lesions and above, suggesting the importance of the splanchnic bed, and possibly arm, in vasoconstriction secondary to postural change.

Recently, direct evidence of the importance of sympathetic regulation has come from the elegant microneurography work of Wallin and and his colleagues. Wallin and Sundlof (1982) showed that in vasovagal syncope there is a cessation of sympathetic outflow and that this was a contributing factor in the hypotension. Their important conclusion was that neurally mediated muscular vasodilatation was due to inhibition of normal sympathetic vasoconstrictor outflow. Wallin and Nerhed (1982) investigated the relationship between sympathetic outflow, blood pressure, and heart rate. The outflow in the peroneal nerve was correlated with cardioacceleration and rises in blood pressure. The same baroreceptor events related to both; i.e. blood pressure alterations were found to be achieved by changes in heart rate and vasoconstrictor outflow via a central control. In tetraplegics this central control may still be present and can affect heart rate via the vagus, but sympathetic outflow will no longer be under descending control since its efferents in the thoracic and lumbar spine are below the lesion. In a further and important study in tetraplegics, Wallin and Stjernberg (1984) found a reduced or absent tonic sympathetic discharge in a nerve below the lesion, with no response to provocative stimuli delivered above the lesion and no response to changes in ambient temperature.

There is a predictable response in heart rate with a fall in blood pressure in tetraplegics. The rate does not usually rise, however, above 100 beats/min. Presumably the withdrawal of vagal tone is normally accompanied by increased sympathetic stimulation, something absent in tetraplegic patients (Mathias and Frankel 1983a). The inability of vagal inhibition to increase blood pressure by increasing heart rate in this circumstance is reduced further by a reduced ventricular pre-load due to poor return of blood to the heart (Rowell 1984).

Despite these considerations it is a widespread clinical experience that, with time after spinal injury, the symptoms of orthostatic hypotension are less and that most patients can adapt to the sitting position and even to a standing-frame reasonably well. How this is achieved, and why they do not suffer from vasovagal syncope, is still not completely clear.

The most obvious way is to reduce or abolish orthostatic hypotension itself, something suggested by Guttman (1976) and Figoni (1984), amongst others. Jonason (1947) showed in two patients with T3 and T4 lesions that over a four-month period post-injury orthostatic hypotension, which had been severe initially, was nearly abolished (Fig. 9.1). Johnson, Crampton-

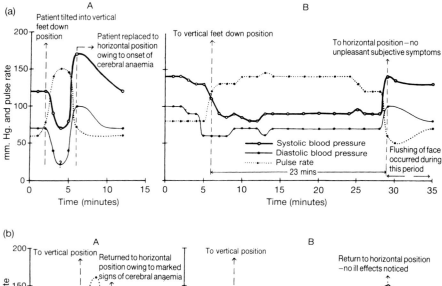

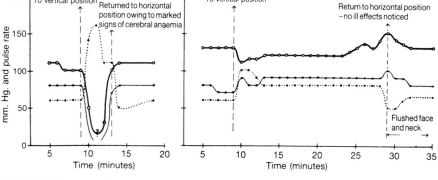

Fig. 9.1. Recovery from postural hypotension in two patients with thoracic lesions (completeness not stated). In each case the vertical head-up position was reached from horizontal in 1.5 s. No mention was made of the rehabilitation between experiments. (a) T4 lesions (A, 30 May 1946; B, 18 September 1946); (b) T3 lesion (A, 30 August 1946; B, 29 October 1946). [Reproduced with permission from Johnason (1947).]

Smith, and Spalding (1969) found in tetraplegics three years post-injury that with repeated tilts over a comparatively short time (40 min), orthostatic hypotension was reduced from 122/85 mm Hg to 92/62 mm Hg initially, and to 100/95 mm Hg at the end of the experiment. There was no significant change in blood volume during this period. The study has been criticized since free urine drainage was not allowed and so some of the effect could have been secondary to spinal reflex vasoconstriction in response to bladder afferents. It is also interesting that these patients, three years after their injury and fully rehabilitated, still showed a postural drop in their pressure. Corbett, Frankel, and Harris (1971c) in their tilt studies found some evidence of a training effect. However, in a study primarily concerned in studying renin release, Mathias, Christensen, Corbett, Frankel, Goodwin, and Peart (1975) investigated four tetraplegics, one of whom showed no hypotension and was five years post-injury. The other three were less than three months since injury and all showed orthostatic hypotension.

Against this body of work, some workers have found no reduction in orthostatic hypotension with time or training. The striking observation from the studies of Corbett et al. (1971c) was that the mean blood pressures fell from 68 mm Hg to 44 mm Hg with a 30° tilt and that the pressures fell as low as 28/15 mm Hg in some patients (Fig. 9.2). This was despite the patients

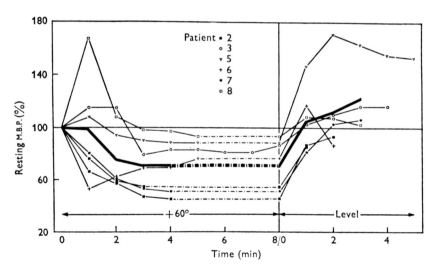

Fig. 9.2. The effect in six tetraplegic patients of seven tilts of +60° on mean blood pressure (MBP), expressed as a percentage of the average value for the three-minute period preceding tilt. The average for all tilts is shown by the heavy line. The patients were C5–7 clinically complete lesions and 4.5–156 (average 67) months post-injury. They all spent much of their days in a wheel-chair. [Reproduced with permission from Corbett et al. (1971c).]

Table 9.1. Resting and minimum mean blood pressure (MBP, mm Hg) and resting and maximum heart rate (HR, beats/min) for tilts of 30°, 45°, and 60° in ten tetraplegic patients (C5–7 complete, a mean of sixty-seven months post-injury, and wheel-chair bound). Note the variation between individuals and within patient 10 in successive trials *a*, *b*, and *c*. [Adapted with permission from Corbett *et al.* (1971c).]

		+30°				+45°				+60°			
		MBP		HR		MBP		HR		MBP		HR	
Patient	Investigation	B[1]	D[2]	B	D	B	D	B	D	B	D	B	D
1		71	31	68	90	70	41	68	113	—	—	—	—
2		79	72	54	73	96	35	49	102	74	31	59	96
3		72	40	57	80	70	49	51	80	70	55	51	92
4		83	53	70	93	79	35	69	105	—	—	62	128
5		56	37	69	83	55	22	72	84	49	43	72	100
6		63	27	106	130	74	19	89	130	74	37	89	125
7		70	45	69	90	90	45	64	96	90	45	64	108
8		49	49	41	44	60	40	35	62	60	50	35	70
9		—	—	—	—	—	—	—	—	86	58	130	180
10	*a*	—	—	—	—	88	26	73	96	—	—	—	—
10	*b*	—	—	—	—	70	45	68	94	—	—	—	—
10	*c*	—	—	—	—	95	44	95	124	—	—	—	—
Average for patients 1–8		68	44	67	85	74	36	62	97	70	44	62	103

[1] Mean value for previous three minutes (B).
[2] Minimum MBP during tilt, and simultaneously measured HR (D).

being an average of sixty-seven months post-injury and being fully re-habilitated. Engel and Hildebrandt (1976) studied a series of tetraplegic patients sequentially over forty-six days with tilts of 20–80° every day and found no significant training effect.

Thus the experimental evidence of a reduction in orthostatic hypotension with time and training is surprisingly less secure than one would expect from the clinical experience. A further contributing factor to the different results is the individual results found in different subjects. In the Corbett *et al.* (1971c) study there was a wide range of falls in blood pressure with posture (Table 9.1). Possible reasons for these conflicting results will be considered later. The mechanisms underlying the presumed recovery from hypotension have been investigated by many groups of workers.

That increased generalized sympathetic activity was involved was excluded by the finding of no detectable rise in circulating catecholamine in response to tilting in tetraplegia (Mathias *et al.* 1975). However, the possibility that some nervous activity was occurring without systemic 'leakage' of neurotransmitter from synapses into the systemic circulation could not be excluded.

Since the work of Carmichael's group in the 1930s, it has been known that in normal subjects following a simple provocative stimulus like a cough a transient wave of vasoconstriction spreads out into the limbs and may be detected in the fingers and toes (see Uprus, Gaylor, Williams, and Carmichael 1935). The technique was first used in spinal cord injured patients recovered from spinal shock by Gilliatt, Guttman, and Whitteridge (1947). They used, as had the earlier study, whole finger plethysmography and measured the vasoconstriction in response to a single deep breath and cough. They showed that these reflexes were still present in high cervical lesions in both fingers and toes. Not only had the reflexes been shown to be spinal, but, of equal or greater importance, co-ordinated spinal reflex activity in the vasomotor system below the lesion had been established. Their technique did not distinguish arterial from venous vasoconstriction, although subsequent work has shown that both do actually exist in these circumstances (Corbett, Frankel, and Harris 1971a). Cole, Mani, and Sedgwick (1985) have repeated the work of Gilliatt *et al.* with the much simpler laser Doppler flowmeter, a device which allows the qualitative measurement of cutaneous blood-flow. They also found that cutaneous vasomotor reflexes were present in chronic tetraplegics, and that the responses were exaggerated in the fingers after cough and deep inspiration, despite these procedures being poorly performed (Fig. 9.3). The reflexes were absent or equivocal in patients with T3–8 lesions, suggesting that the upper thoracic cord is necessary for their integration. Both Gilliatt *et al.* and Cole *et al.* found flexor spasm of the foot or just a small spontaneous flexion movement of the big toe to be powerful stimuli for vasomotor reflexes, or at least for associated events with such reflexes.

These reflexes, therefore, can be demonstrated in chronic tetraplegics. Do they make significant contributions to the recovery from orthostatic hypotension in such patients? Corbett *et al.* (1971a) measured pulse blood pressure, hand blood-flow, and occluded vein pressure in tetraplegics in response to spontaneous and induced flexor spasms of the foot. They found that the stimuli led to an increase in systolic blood pressure with a reduction in heart rate. The former they thought was due to an increase in peripheral resistance, with both arterial and capacitance vessels contributing. However, the mean rise in mean blood pressure in five patients was from 67 to 80 mm Hg and lasted around 20 s. In their second paper (Corbett, Frankel, and Harris 1971b) they investigated the effect of bladder percussion, pinprick, and cold packs applied below the lesion upon their measured variables. Bladder percussion elevated mean blood pressure from a mean of 88 mm Hg to 103 mm Hg, the other stimuli having only a very

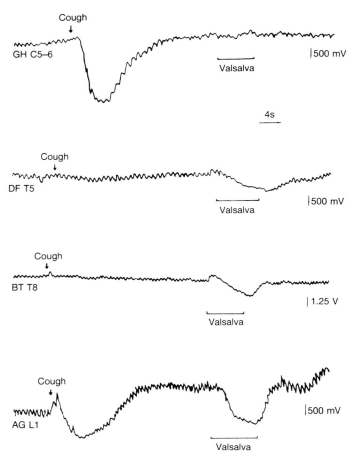

Fig. 9.3. Cutaneous vasoconstrictor reflexes in the thumbs of four patients with different levels of complete spinal cord lesion. They were asked to cough once, or blow into a mercury column to a pressure of 20–40 mm Hg for ten seconds.
[Reproduced with permission from Cole, Mani, and Sedgwick (unpub.).]

small effect (Fig. 9.4). They commented that the fact that an ice-pack on the abdomen had an effect might be surprising given the defect in thermoregulation in tetraplegics. It is less so if it is considered as a painful stimulus instead.

It is therefore theoretically possible that these reflexes may be physiologically useful, but they occur so infrequently, so unpredictable, and so transiently that this appears unlikely. Corbett *et al.* (1971*b*) concluded that the effects they had observed were so inconsistent that to even call them reflexes was uncalled for. However, at a discussion of their work at a contemporary meeting Rossier described a patient who had learnt to use

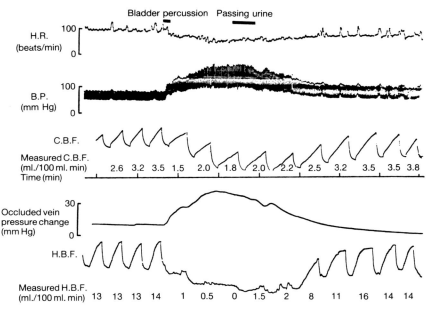

Fig. 9.4. The effect of bladder percussion on various cardiovascular parameters in a rehabilitated tetraplegic patient. Bladder percussion led to an elevation of blood pressure for three minutes or so in a case in whom the bladder was full enough to void. HR, heart rate; CBF, calf blood-flow; BP, blood pressure; HBF, hand blood-flow. [Reproduced with permission from Corbett *et al.* (1971*b*).]

bladder percussion when he felt faint when upright, so under some circumstances they may be useful (Corbett, Frankel, and Harris 1971*d*). This also raises the question of the relation between these reflexes and autonomic dysreflexiae, something that will be considered later. Corbett *et al.* (1971*b*) also found a greater reduction in blood-flow to the hand than to the foot. If the two methods were equally accurate in the two regions this implies a greater autonomic reflex control over the arm blood-flow than the leg (see Guttman and Whitteridge 1947) or that these reflexes are hyperactive in the hands of tetraplegics.

In addition to peripheral reflexes there is also some evidence of vasomotor reflexes occurring in the splanchnic bed. This was considered by Cunningham, Guttman, Whitteridge, and Wyndham (1953), who suggested that the responses of the circulation in the arm to bladder percussion in high lesions might parallel that in the visceral organs since their autonomic supply comes from similar and overlapping cord levels. Beacham and Kunze (1969) found that some renal afferents increased their firing rates in response to small rises in renal vein and ureteric pressure, and that this reflex was still present in spinal cats. Andrews, Andrews, and Orbach (1971) found in animal studies that an increase in mesenteric venous pressure led to an increase in efferent activity in nerves to the gut, which was still present after C7 spinal cord transection. Krebs, Ragnarrson, and Tuckman (1983) have shown that renal blood-flow and renal vascular resistance altered in response to tilt in both normal and spinal man. They suggested that even in the absence of supraspinal control the renal vasculature could respond to hypotension with some vasoconstriction via a presumed spinal reflex. The quantitative importance of this effect has not been established, but splanchnic vasomotor reflexes may be useful in tetraplegics.

There is some evidence that circulatory reflexes can occur independently of the cord. Gaskell and Burton (1953) [see Rowell's (1984) review] suggested that venous distension could lead to constriction of the arterioles that supply them. This local sympathetic venoarteriolar 'reflex' would presumably be independent of co-ordinated spinal activity. Henrikson (1977) gained some evidence for such an effect in normal subjects with arteriolar reflex constriction in adipose tissue elicited by an increase in venous pressure of 25 mm Hg or more. This was not influenced by acute spinal sympathetic block. In a later study it was shown in normal subjects that blockade of this reflex in one leg (by intra-arterial phentolamine) reduced the normal local venoarteriolar reflex by about 50 per cent (Henrikson, Skagen, Haxholdt, and Dyrberg 1983). This group has also studied a group of twelve tetraplegics and again measured subcutaneous blood flow with various tilts (Skagen, Jensen, Henrikson, and Knydsen 1982). With head-up tilt, subcutaneous flow to the leg fell by 47 per cent with no change after proximal nerve blockade. With leg-down tilt alone, flow was also reduced by a similar amount, even though systemic blood pressure and pulse rate did not alter. Local infiltration of lignocaine into the area, from which xenon wash-out was measured, at doses insufficient to abolish myogenic activity but enough to block sympathetic activity abolished the response. In the arm the situation appeared different. Head-up tilt led to a fall in blood-flow by 37 per cent, but proximal nerve blockade did abolish the response, suggesting that spinal outflow was in this case responsible.

These elegant experiments have demonstrated that local mechanisms in the leg can increase vascular resistance with tilt. However, during these experiments orthostatic hypotension still occurred. The local reflexes may contribute to recovery from the effects of tilt, but they do not abolish them.

Despite the evidence reviewed above for neurological mechanisms which may underlie the reduced effects of postural hypertension with time, in most of the experiments some hypotension still occurred. This has led several groups to investigate whether or not hormonal mechanisms may contribute to the recovery of some stability of blood pressure with tilt in tetraplegics.

Johnson, Park, and Frankel (1971) infused angiotensin into tetraplegic patients during tilt to 45° and observed some recovery in blood pressure over several minutes, a difficult study since this effect is known to occur spontaneously. The patients' time since injury was not stated and, although not given, the doses of angiotensin were said to be physiological. The study suggested that some of the recovery from orthostatic hypotension may occur through the renin–angiotensin system. The authors were careful to stress that the hormonal effect was not sufficient to maintain blood pressure at pre-tilt levels in patients.

In order to suggest that hormonal factors may be important it is also necessary to show that hormone levels rise in the appropriate situation. Mathias *et al.* (1975) showed a rise in renin and then aldosterone in a study of four tetraplegics with tilt. Three of the subjects were, however, less than eight months post-injury, whilst the latter patient was sixty-two months, and it was the latter that had the smallest fall in blood pressure and also the smallest rise in renin—perhaps not the expectation if recovery depended on renin release. Second, there was no direct relation between renin rise and aldosterone level; rather, the latter appeared related both to renin and possibly to the postural fall in blood pressure. As Mathias *et al.* (1975) pointed out, aldosterone levels may reflect not only renin but also such factors as splanchnic circulation and hepatic clearance. However, the study did show the potential availability of hormonal mechanisms to assist recovery from postural hypotension. In later studies, Mathias, Christensen, Frankel, and Peart (1980) have shown that renin release is independent of sympathetic activity and that angiotensin is likely to act directly on renal musculature (Frankel, Mathias, Peart, and Unwin 1981).

All the work on the possible mechanisms of recovery from orthostatic hypotension has been done, necessarily, in acute experiments. However, recovery takes place over weeks in a gradual way, with slow increases in the duration of tilting, for instance. If one postulates that chronic training acts through different mechanisms to those in acute studies then some of the conflicting results in this field may be reconciled. Johnson, Crampton-Smith, and Spalding (1969) found an effect over the short period of forty minutes, but their patients were four years post-injury and so presumably were rehabilitated. In Engels and Hildebrandt's (1976) study and in most others the actual training, in terms of how long each day it occurred, was not given.

It may be the case that effects occur gradually as postural training is increased and so different results would occur if the patients have received postural training for different lengths of time, different durations, and different lapses of time since the injury. Results from acute experiments might be different from those obtained from more chronic ones. This approach, if correct, may give a clue to a possible mechanism of recovery, and has been considered by Mathias (pers. comm.).

It is a well-observed clinical phenomenon that the urine output of a tetraplegic patient is increased on lying down, probably because renal perfusion pressure is increased. It is possible that by lying flat tetraplegics may become relatively volume-depleted, and that this in itself may worsen postural falls in blood pressure. As more upright postures are adopted for longer, plasma volume may become expanded and the postural hypotension be reduced. Such an effect might occur through hormonal factors [both antidiuretic hormone, secretion of which is influenced by posture (Anger, Zehr, Siekert, and Segar 1970) as well as the renin system] and by more direct effects on renal blood-flow. Such a theory would suggest that resting aldosterone levels would be higher in chronic tetraplegics compared with acute patients lying flat. In work conducted thus far there is no evidence for this; in fact, resting aldosterone levels in chronic tetraplegics are lower (Claus-Walker, Carter, Lipscomb, and Vallbona 1968, 1969; Mathias et al., 1975). Claus-Walker et al. found that the postural regulation of aldosterone release in tetraplegics was absent, but interestingly its salt-regulating effect, and hence an effect on plasma volume, was still intact. It may be worthwhile to investigate a group of recently injured tetraplegics prospectively through their rehabilitation periods and measure such variables as response of blood pressure to tilt, relating it carefully to time spent upright, plasma volume, and levels of renal hormones to support or repudiate such theories.

In conclusion, tetraplegics recover in time and with training from postural hypotension and there is good evidence that some of this effect is due to a lessening in the hypotension itself. Mechanisms by which this may be achieved involve local venoarteriolar reflexes and longer term hormonal effects. Co-ordinated spinal vasomotor reflexes may also have a role, with those involving the renal and splanchnic beds being more important than the peripheral ones (see Cunningham et al. 1953). However, Meinecke et al. (1971) did report that postural retraining was worse in high thoracic tetraplegics than in cervical ones, implying possibly that some thoracic sympathetic outflow was of use in this situation. However, it remains the case that the mechanisms of recovery from orthostatic hypotension are poorly understood. Care has thus far been taken to distinguish recovery from the effects of orthostatic hypotension and recovery from the hypotension itself. There is one mechanism by which the former may be achieved without the latter—if cerebral blood-flow autoregulation alters in tetraplegia.

The possible sympathetic contribution to cerebral blood-flow regulation
Cerebral arteries down to and including pial arterioles have a rich sympathetic supply (and parasympathetic supply) (for a review, see Edvinsson and MacKenzie 1977). Whilst it is well known that cerebral autoregula-

tion of blood-flow occurs over a wide range of blood pressures, might the autonomic system assist at each end of that range?

The lower limit of regulation in the baboon has been shown to be about 65 mm Hg in controlled haemorrhage experiments. This falls to about 35 mm Hg after an acute cervical sympathectomy, with the implication that sympathetic discharge limits the ability of cerebral arteries to dilate at low arterial pressure. Corbett *et al.* (1971*c*) found that on tilting tetraplegics, blood pressures of 35/20 mm Hg occurred apparently asymptomatically and in one case a level of 29/19 mm Hg occurred when syncope resulted. These are pressures lower than those found in normal man (Bevan, Honour, and Stott 1969). It is indirect evidence that co-ordinated sympathetic activity in control of cerebral blood-flow is absent in tetraplegics. Since these patients can withstand pressures lower than a normal population, it also suggests that, as in the baboon, removal of sympathetic activity may allow auto-regulation at lower pressures. On the other hand, stimulation of the cervical sympathetic trunk has been shown to lead to cerebral arterial constriction in baboons (Harper, Deshmukh, Rowan, and Jennett 1972). There have been two main studies on cerebral blood-flow in tetraplegia.

Eidelmann (1973, and quoted in Mathias and Frankel 1983a) found, as one might expect, that tetraplegics autoregulate at lower perfusion pressures. This may be due to two effects. First, as discussed above, the withdrawal of sympathetic activity may affect the extreme lower end of autoregulation. Second, tetraplegics have lower resting blood pressures than normals and this may lead to alterations in cerebral blood vessel tone and so shift autoregulation also, by venoarteriolar reflexes and by effects on vessel tone itself (see Sharpey-Schaffer 1961). Certainly, after induced chronic hypertension in baboons, cerebral autoregulation breaks down at higher arterial pressures with acute hypotension, suggesting that such shifts do occur at least at the top end (Jones, Fitch, MacKenzie, Strandgard, and Harper 1976). Eidelman (1973) was constrained from looking at really high and low pressures by ethical considerations. His study has also been criticized on methodological grounds. A xenon wash-out technique was employed and flow through extracranial vessels may have contaminated his results. However, this seems insufficient to seriously undermine his conclusions. In the other study, Nanda, Wyper, Harper, and Johnson (1974) found the response of the cerebral circulation to both hypotension and hypocapnia to be normal in tetraplegics and therefore to be independent of sympathetic control. However, in this study the blood pressure only fell to a mean of 70 mm Hg, a level higher than that at which the sympathetic system would be physiologically active.

Thus there is evidence that tetraplegics autoregulate at lower pressures, and indirect evidence that the removal of the sympathetic supply may contribute to this effect at low pressures (Fig. 9.5). Mention should also be

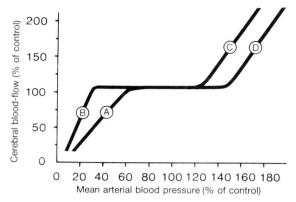

Fig. 9.5. The effect of sympathetic nervous activity on cerebral autoregulation of blood-flow: a schematized graph. (A) Lower limit of autoregulation during haemorrhage; i.e. accompanied by a sympathoadrenal response. (B) Lower limit of autoregulation during haemorrhage with either adrenergic blockade or cervical sympathectomy. (C) The upper limit of autoregulation to drug-indiced hypertension. (D) The upper limit of autoregulation to drug-induced hypertension with stimulation of the cervical sympathetic trunk. Note that the scales are percentages. [Reproduced from Edvinsson and MacKenzie (1977) by kind permission of the American Society for Pharmacology and Experimental Therapeutics.]

made of the effect on the cerebral response to high perfusion pressures in tetraplegic patients deprived of co-ordinated sympathetic afferents.

Bill and Linder (1976) found that stimulation of the cervical sympathetic trunk during induced hypertension moved the upper limit of cerebral autoregulation to higher absolute levels of arterial pressure. This suggests that in tetraplegics the upper limit of autoregulation may be lower than in normals, who can experience high blood pressures during some circumstances (Bevan *et al.* 1969).

In addition to effects on cerebral blood pressure there is also some evidence of sympathetic involvement in the regulation of cerebrospinal fluid production. The mean production falls by about 30 per cent after stimulation of the cervical sympathetic trunk, and rises by a similar amount compared with normal control animals after excision of the trunk (for a review, see Edvinsson 1982). Whether this effect also occurs in man and whether or not it is significant in those with spinal injuries has yet to be determined.

Tetraplegic patients may, therefore, be at a compounded risk from the effects of paroxysmal hypertension because of three factors. First, they tend to live at lower blood pressures and may therefore have shifted their limits for autoregulation downwards to these lower perfusion pressures. Second, they are without sympathetic supply and so cannot extend their limits upwards in the face of high blood pressures. Last, they have to face higher blood pressures than normal subjects, more often, more extremely, and for longer durations. It is this last factor, autonomic dysreflexia, which will be considered next.

Autonomic dysreflexiae

The symptoms of autonomic dysreflexia were first observed by Hilton (1860) and subsequently studied in their classic papers by Head and Riddoch (1917) and by Guttman and Whitteridge (1947) (see above).

The clinical symptoms are of abdominal discomfort followed by an ill-defined ascending feeling, oppression in the chest, and parathesiae around the neck and shoulders. Later, difficulty in breathing may occur with shivering or a feeling of being hot (Guttman 1976). Unfortunately, neuro-logical symptoms may also be the presentation, with aphasia, blindness, and convulsions. The signs are of severe hypertension, reaching 255/135 mm Hg in one of Guttman and Whitteridge's (1947) patients (Fig. 9.6), usually a bradycardia, and pupillary dilatation (for an account of the latter, see Wayne and Vukor 1977). Above the lesion there can be initial pallor, followed by flushing of the face and neck with severe sweating. Below the lesion the peripheries are cold.

The incidence of autonomic dysreflexia depends on its definition. Lindan, Joiner, Freehafer, and Hazel (1980) defined it as being a sudden rise in both

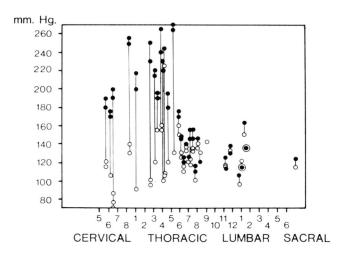

Fig. 9.6. The relation between blood pressure change and the level of complete spinal cord lesion following bladder distension. Open circles, systolic blood pressure before bladder distension; closed circles, systolic blood pressure during bladder distension. [Reproduced with permission from Guttmann and Whitteridge (1947).]

systolic and diastolic pressures, with or without headache, and with accompanying symptoms of sweating, pupillary change, and skin blotching. Any rise in blood pressure of more than 10 mm Hg during cystometry was considered to mean that dysreflexiae were present. Forty-eight per cent of 213 patients were thought to have dysreflexiae. Most clinicians adopt a more symptomatic approach. Kewalramani (1980) suggesting a figure nearer 30 per cent of complete lesions above T5. This author found it to be much more common in males (9:1), in disagreement with Lindan *et al.* (1980), and commoner in younger patients in whom autonomic functions may be more active (although, unfortunately, he did not give the age-distribution of his whole sample). It tends to occur about six months post-injury and has occurred in 92 per cent of those who will have it by a year. In seventy-eight out of eighty-eight of the sample used by Lindan *et al.* it presented with headache and sweating.

The stimuli to provoke autonomic dysreflexiae have been known to be dilatation of and abnormal activity in the urinary bladder, and to a lesser extent the rectum, since some of the earliest descriptions. Amongst the other causes have been birth (Guttman 1963; McGregor and Meeuwsen 1985), electrical ejaculation (Guttman and Walsh 1971), in-growing toe-nails (Corbett *et al.* 1971a), and even passive stretching of joints (McGarry, Woolsey, and Thompson 1982). It is possible that all patients with injuries above T6 would develop dysreflexiae under the right provoking circumstances. Why some patients are more prone to them is not known. It is likely that better post-injury care and prevention where possible of situations in which dysreflexiae could occur has reduced their seriousness to an extent. Unfortunately, there are no predictive clues as to which patients are most at risk.

In those followed up only 8 per cent were free after a year, whilst in 48 per cent they persisted for six years and in 17 per cent for up to twenty years post-injury. The symptoms are personally distressing and socially embarrassing. It also carries a morbidity from cerebrovascular events and a not inconsiderable mortality. Its presence is difficult to predict, although Kewalramani (1980) reported that, as long as the bulbocavernosus reflex was absent (an index of spinal shock), dysreflexiae did not occur.

The pathophysiology was first investigated by Guttman and Whitteridge (1947). They showed a rise in both systolic and diastolic blood pressure, with an increased occluded venous pressure and a reduction in peripheral blood-flow. Cardiac output, stroke volume, systemic vascular resistance, and pulmonary arterial pressure also rise (Corbett, Debarge, Frankel, and Mathias 1975). It is likely that the vasoconstriction in the splanchnic bed is a large contributory factor, since dysreflexiae are seen in those patients with lesions above the splanchnic sympathetic outflow and since this bed contains a large percentage of the cardiac output (estimated by Krebs *et al.* 1983 to be

50 per cent). That the rise in blood pressure was greater than expected from the rise in cardiac output was shown by Cunningham, Guttman, Whitteridge, and Wyndham (1953). They suggested that a large component was due to splanchnic vasoconstriction. They also cautioned against considering that the circulation to the arm and leg has similar vasomotor control in all circumstances. All these vasoconstriction effects may be exaggerated since, in the tetraplegic, sympathetic tonic vasoconstrictor outflow is severely reduced (Wallin and Stjernberg 1984).

Guttman and Whitteridge (1947) considered whether adrenalin release might cause the effects they observed. They thought it unlikely for several reasons. For example, in patients with mid-thoracic lesions and higher, vasodilatation occurred in the fingers and head, respectively. If adrenalin levels had been high this would not have been expected. More recently it has been possible to measure levels of circulating catecholamines and it has been shown that during autonomic dysreflexiae neither adrenalin nor renin rises (see Fig. 9.7; Mathias, Christensen, Corbett, Frankel, and Spalding 1976; Mathias, Frankel, Davies, James, and Peart 1981). Noradrenalin did rise, and since this is released from sympathetic terminals this supports an

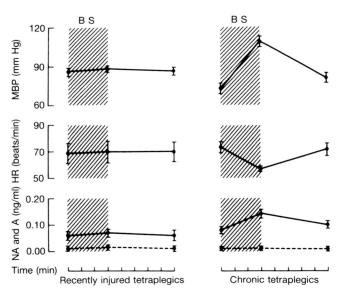

Fig. 9.7. The average levels of mean blood pressure (MBP), heart rate (HR), and plasma noradrenalin (NA continuous line) and adrenalin (A discontinuous line) in recently injured and chronic tetraplegics before, during, and after bladder stimulation (BS). The bars represent standard errors of the mean, ±. In the recently injured group, no changes occurred. In the chronic group, MBP and NA rose, and HR fell. No change in A was seen. [Reproduced with permission from Mathias *et al.* (1979*a*).]

overactivity in the sympathetic system if not one involving increased and co-ordinated adrenal output.

Head and Riddoch (1917) commented that an important part of the syndrome was that relatively small stimuli below the lesion no longer elicited reflexes within their usual influence, but rather spread throughout the cord caudal to the damage, their term being the mass reflex. Frankel and Mathias (1979) commented that the loss of supraspinal control was obviously important and wondered whether these reflexes are present normally but inhibited and only unmasked in tetraplegia. Possible roles were suggested to be in postural adaptations of blood pressure and in maintaining perfusion of viscera during their distension. However, it is odd that orthostatic hypotension is not usually a trigger for a dysreflexia. The possible physiological mechanisms underlying this phenomenon will be considered in the discussion.

If autonomic dysreflexiae represent a sympathetic storm in the absence of descending control and only partial parasympathetic compensation via vagally mediated bradycardia, there is one situation in which a 'parasympathetic storm' might be said to occur. In patients with high cervical lesions involving segments supplying the phrenic nerve, ventilatory

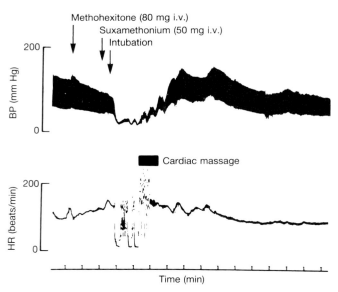

Fig. 9.8. The effect of endotracheal intubation on blood pressure (BP) and heart rate (HR) in a chronic tetraplegic patient being anaesthetized for surgery. A temporary and reversed cardiac arrest occurred. [Reproduced from Welply *et al.* (1975) by kind permission of Churchill Livingstone, Edinburgh.]

assistance is often required for varying lengths of time. In this situation tracheal irritation, e.g. suction, can produce vagal overactivity with profound bradycardia and even cardiac arrest (see Fig. 9.8; Dollfus and Frankel 1965; Frankel, Mathias, and Spalding 1975; Welply, Mathias, and Frankel 1975). Avoidance of hypoxia and the use of atropine should be considered in these cases.

The most important clinical *sequelae* of the loss of autonomic co-ordination in spinal cord injury involve the cardiovascular system, and so these have been discussed in some detail. Some other aspects will now be considered.

Thermoregulation

The consequences of spinal injury for thermoregulation have been known for a hundred years (see above). Both hypo- and hyperthermia are apparently more likely immediately after spinal cord injury than later (Guttman 1976), with the implication that some recovery in this occurs with time, although the mechanisms for this have not been investigated. Pledger (1962) showed that it was important to monitor central temperature, since oral readings may remain normal when, say, rectal ones are low.

Detailed measurement of the response of patients to high and low ambient temperatures were made by Guttman, Silver, and Wyndham (1958). In patients with lesions above T6 there was little thermoregulatory response to changing temperature, because of a loss of shivering, no thermoregulatory sweating, and no peripheral circulatory adjustments below the lesion. Johnson (1971), however, found some pilo-erection, vasoconstriction, and vasodilatation below complete lesions, the mechanism being thought to be spinal cord reflex activity. Guttman, Randall, and Silver (quoted in Guttman 1976) also found that in response to intense heat small patches of sweat appeared on the skin below the lesion. Guttman (1976) distinguished these areas from the normal areas of sweating above the lesion.

In a thorough reinvestigation, Normell (1974) employed thermography and plastic sheeting (to assess sweat loss). He found no significant spinal thermoregulatory function below complete lesions. Once more, as in the case of peripheral vasomotor function, spinal reflexes have been shown, but their physiological usefulness has not been established.

Gastrointestinal and metabolic *sequelae*

The consequences of spinal cord injury upon large bowel function have been apparent clinically since such patients' survival periods have been prolonged. More recently the upper gastrointestinal tract has been studied more, not least because of the high incidence of acute bleeds after injury. Kewalramani (1980) found that 5 per cent of a sample of 576 new patients had acute bleeds, with 23/24 being in those with lesions above T7, and 21 of those with C4–6 lesions. It is thought to occur because of increased vagal activity leading to stomach acidity (Pollock, Boshes, Chor, Finkelman, Arieff, and Brown 1951; also Pollock and Finkelman 1954, quoted in Kewalramani 1979), although high gastrin levels may also be involved (see Mathias and Frankel 1983a). The pathogenesis may be seen to be more complex still as our understanding increases. The higher incidence in cervical lesions may reflect that these are more severe injuries, but more likely that ulceration is not due solely to vagal action but rather to vagal activity in the absence of any co-ordinated sympathetic actions of an as yet poorly understood nature. It is of interest that whilst visceral abdominal pain is reduced in spinal injury, so that the classical signs of ulceration and perforation may be absent, visceral sensations of hunger are still present, suggesting that their afferent pathway involves the vagus (Crawford and Frankel 1971). More recently, hepatic glucoreceptors and intestinal chemoreceptors have been described which may reduce appetite in response to certain amino-acids (Niijima and Mei 1984).

Paralytic ileus and acute gastric dilatation may also occur with spinal shock and later (Watson 1981). The mechanisms for these events remain unclear since the vagus is intact. As Guttman (1976) implied, the loss of sympathetic effects may play a part. It is probably an oversimplification to consider the parasympathetic system to function normally but to be hyperactive in the absence of co-ordinated sympathetic activity. The two are unlikely to be simple antagonists in intact man.

The loss of sacral parasympathetic regulation leads to atony of the large bowel and retention of faeces. Connell, Frankel, and Guttman (1963) showed that the behaviour of the rectum after spinal injury depended on the level of the cord lesion and the reflex state. With cervical lesions there was a reduction in motility, whilst in lumbar ones increased colonic activity occurred. This hyperactivity was abolished by intrathecal alcohol, suggesting an autonomy of control over the large bowel through the remaining cord, with the lumbar segments inhibiting a possible 'sacral parasympathetic centre'. The ingestion of food increased activity in the large bowel in cervical as well as in lumbar injuries, but with a latency of up to fifteen minutes so that it is not necessary to invoke a neural mechanism. The response of the sigmoid to rectal distension also depended on the lesion's level. In cervical ones there was an initial inhibition followed by increased activity, as well as in some cases signs of autonomic dysreflexiae. In flaccid low lesions there was no response in distension. The reflexes may be homologous to cutaneous vasomotor reflexes and require an intact cord below the lesion for their functioning. However, unlike the cutaneous reflexes, some large bowel function and some reflex evacuation of faeces may develop and be useful, this being compatible with Guttman's (1976) idea that regulation of these functions could be mediated through a low spinal cord 'centre' independent of higher control.

More recently, Andersson, Bloom, Edwards, Jarhult, and Mellander (1983) in animal studies have found that stimulation of pelvic nerves led to a maintained rectal vasodilatation, but a more transient one in the colon, and that muscarinic block reduced the latter effect but not the former. They implicated vasoactive intestinal polypeptide. They therefore demonstrated that differential effects on different parts of the large bowel may occur through pelvic nerves.

Despite these studies the functioning of the bowel has received little attention, since clinically it is little affected in terms of its absorption functions in tetraplegia. It is the excretory functions which have been considered to be more affected. There is the beginning, however, of a realization that absorptive and metabolic consequences of injury may be demonstrated.

The consequences of insulin-induced hypoglycaemia were investigated by two groups in 1979. Mathias, Frankel, Turner, and Christensen (1979b) infused fish insulin into tetraplegics and found that, during hypoglycaemia, hypotension occurred without the normal responses of rises in adrenalin and with a suppression of endogenous insulin (Fig. 9.9). The fall in blood pressure was thought to be due to an effect of insulin in reducing intravascular volume, with once more an inability to mount a co-ordinated sympathetic response. The suppression of endogenous insulin was suggested to be a direct effect of the hypoglycaemia on end-organs and to be independent of sympathetic neurons. An important point clinically was that some of the clinical symptoms of hypoglycaemia were absent in these patients, and so hypoglycaemia, like hypothermia and gastrointestinal haemorrhage, may be missed in these patients.

The metabolic effects were studied by Corrall, Frier, McClemont, Taylor, and Christie (1979). In hypoglycaemia, tetraplegics were found to have smaller rises in blood lactate and their levels of free fatty acids fell more and later than in normals. This delay in glycogenolysis was ascribed to a lack of adrenal response. The situation may be more complicated; in work using the more recently available hormonal assays, Frier, Corrall, Adrian, Bloom, Strachan, and Heading (1984) found that after a standard meal tetraplegics showed a mild glucose intolerance. They were unable to explain this simply in terms of abnormal insulin or glucagon secretion, and concluded that there may be a sympathetic inhibitory influence on the secretion of pancreatic polypeptide. With the recent advances in knowledge of the number and diversity of functions of gastrointestinal peptides there is much work to be done on the metabolic effects of spinal injury.

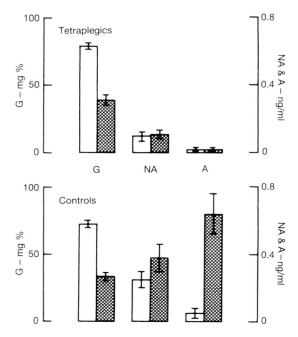

Fig. 9.9. The effect on levels of circulating factors of insulin-induced hypoglycaemia in chronic tetraplegics and controls. G, blood glucose; NA, plasma noradrenalin; A, adrenalin. Before hypoglycaemia: blank histograms; during hypoglycaemia: hatched histograms. In normal subjects the hypoglycaemia led to a small rise in noradrenalin and to a marked one in adrenalin. In tetraplegics no such response was observed. [Reproduced from Mathias *et al.* (1979*b*) by kind permission of Churchill Livingstone, Edinburgh.]

Renal consequences of spinal injury

Work on renal mechanisms in tetraplegia has centred on the renin–angiotensin–aldosterone system and its possible role in blood pressure and blood volume regulation. It has been shown that renin release is independent of sympathetic activity (Frankel, Mathias, and Peart 1980). In animal studies other workers have concluded from acute experiments that renal nerves have no role in renal haemodynamics and excretion (Stella, Golin, Busnardo, and Zanchetti (1984) nor in tubuloglomerular feedback (Hermansson, Kallskog, and Wolgast 1984).

However, in tetraplegia it is the chronic effects which are likely to be of more relevance. In normal human subjects, Kramer, Stinnesbeck, Klautke, Kipnowski, Klingmueller, Glaenzer, and Duesing (1985) studied the effect of a low sodium diet on urinary excretion of various relevant hormones.

Their conclusion was that the renal adrenergic system has a role in controlling renal vascular resistance and in sodium conservation during low sodium intake. That intact renal innervation was required for normal renal sodium conservation was also found from animal experiments by Dibona and Sawin (1983). Osborn, Roman, and Harland (1985), in further animal studies, investigated the effect of low rates of renal nerve stimulation on sodium and bicarbonate excretion. They concluded that nerve activity may be involved in normal regulation of acid–base balance.

In view of these results and those considered previously, implicating renal hormones in long-term cardiovascular homeostasis, it would be of interest to reveal possible deficits in renal function in man after spinal cord injury; for example, the effect of a low sodium diet on renal conservation, blood volume, and orthostatic hypotension.

Bone

Effects of spinal cord injury upon bone metabolism are well known but poorly understood. Osteoporosis and an increased frequency of fractures and, in particular, of calcification occur (see Guttman 1976 for a review). Hernandez, Former, de la Fuente, Gonzalez, and Miro (1978) surveyed 704 tetra- and paraplegic patients and found that 20 per cent had ectopic calcification around the hips and that this never appeared above the lesion. Scher (1976) suggested that it was more common in complete lesions.

Griffiths, d'Orsi, and Zimmermann (1972) and Griffiths, Bushieff, and Zimmermann (1976) used an osteodensitometer to investigate the bone of spinal injured patients. They found that while the cortical bone was normal there was a differential loss of trabecular bone both in tetraplegics and, surprisingly, in paraplegics. This pattern of loss is not that seen in idiopathic osteoporosis but is more like that observed in such conditions as reflex sympathetic dystrophy and diabetic amyotrophy, both thought to involve neural dysfunction. They suggested that the changes might be related in altered blood-flow, maybe through arteriovenous fistulae. This has received some support from Chantraine, van Ouwenaller, Hachen, and Schinas (1979), who found that, after spinal injury, intramedullary pressure in bone increased to double that found in normal subjects and that there were signs of delayed blood-flow and diaphyseal reflux. They thought these effects were due to opening of arteriovenous fistulae secondary to vasomotor paralysis. It may be homologous to the loss of usual vasoconstriction to muscle and skin after spinal injury, although how this relates to the resulting osteoporosis is not known. If venous stasis were to result in relative hypoxia this might contribute.

Bone metabolism was studied longitudinally after injury by Bergman, Heilporn, Schoutens, Paternot, and Tricot (1977). They found that modelling was enhanced in the first three months post-injury, especially in the non-paralysed area, with later modelling going on below the lesion. After three months, calcium balance above the lesion was unaltered, but it continued to be negative below it. Griffiths *et al.* (1976) suggested that remodelling and osteoporosis continues for five years after the injury but then appears to be stable. After injury there is increased excretion of calcium, which appears to be independent of diet (Guttman 1976). Alkaline phosphatase is not elevated, and there are no apparent long-term changes post-injury in calcitonin or parathyroid hormone levels (Claus-Walker, Scurry, and Carter 1977).

It is a clinical experience that osteoporosis may be reduced and even partially reversed with postural training and weight-bearing, and also, possibly, with functional electrical stimulation. Kaplan, Roden, Gilbert, Richards, and Goldschmidt (1981) found the postural effect to be greater than the weight-bearing one. The mechanism for this effect has been suggested to be increased blood-flow, although more correctly a better physiological distribution of bone blood-flow may be necessary. It is difficult conceptuality to see how postural change would promote this. Possibly the increased venous pressure in bone vessels secondary to the upright position may activate venoarteriolar reflexes (see Henrikson 1977 and above) and reduce venous stasis and arteriovenous shunts. However, since Chantraine *et al.* (1979) found that intramedullary pressure at rest was twice normal, one must then postulate that pressures greater than that are required to activate these reflexes (whereas in subcutaneous tissue venous pressure increases of 25 mm Hg were sufficient).

Thus the effects upon bone of spinal injury have been well studied, although the neurological, cardiovascular, and endocrinological influences upon the tissue have yet to be well defined.

Pain

Pain is a relatively common occurrence post-injury (Rowe, Robinson, Ells, and Cole, in the press). In one survey 60 per cent of patients had some pain two years post-injury and, in a not inconsiderable number of these, pain was felt at or below the lesion. The cause of this is presumed to be abnormal spinothalamic activity, but is extremely poorly understood, let alone treated. This is mentioned in a review of autonomic effects of spinal cord injury because, very rarely, reflex sympathetic dystrophy can co-exist with spinal cord injury (Andrews and Armitage 1971) and because of the use of sympathectomy in some pain syndromes. This most distressing symptom should not be forgotten, especially when important advances in the knowledge of neurotransmitters in the spinal cord are being made.

Discussion: the pathophysiological basis of autonomic responses after spinal cord injury

Head and Riddoch (1917) considered autonomic dysreflexiae to represent abnormal activity occurring in the sympathetic system, with alterations in reflex activity after recovery from spinal shock. This has the most important implication that the effect of spinal cord injury is more than a simple disinhibition. It will be suggested in this section that autonomic dysreflexiae may be begun to be understood in terms of derangements in synaptic physiology and pharmacology and even in connections within the autonomic system below the lesion which are initiated by the loss of descending control.

Some stimuli in tetraplegics lead to transient and short rises in blood pressure, whilst other more potent ones to dysreflexiae with responses exaggerated in spread duration and intensity. Why do tetraplegics not suffer from these repeatedly? It is as though there is a threshold effect with inputs above a certain intensity leading to the catastrophic event. If one considers that under normal circumstances the potential for this is present but is under strong descending inhibition, one might expect them to occur when the restraint is removed. Yet dysreflexiae tend to occur after six months or so, later than the emergence from arreflexic spinal shock. This suggests that an abnormal organization has occurred. Another way to consider a threshold effect, in which stimuli have to reach a certain level before they produce an effect, is that it may represent a breakdown of a normally functioning self-regulation or negative feedback to prevent such large sympathetic outflows. Is there any evidence for synaptic mechanisms that might underlie some of these postulated changes?

There is considerable evidence from animal work that transmission through autonomic ganglia is not a simple in–out function—or why have a ganglion or a synapse (Prigioni and Casella 1984)? An idea which is beginning to become accepted is that there are at least two types of neurotransmitter, with the classical ones (acetylcholine and noradrenalin) being concerned with fast synaptic action. The more recently investigated neurotransmitters are likely to act as neuromodulators altering neuronal excitability in subtle ways with longer time courses of action (for reviews, see Nicholl 1982; Erulkar 1983; and Horn and Dodd 1983). For example, they may lead to excitatory post-synaptic potentials of low amplitude, but with time courses of several seconds rather than of milliseconds. Over the last twenty years or so, over forty peptides have been discovered with properties of neurotransmitters, including ones previously thought to be hormones; e.g. angiotensin II, vasoactive intestinal peptide (VIP), and luteinizing hormone releasing hormone (LHRH). The autonomic system, with its slow conducting pathways and its often long-term changes in responsiveness, would seem to be a system in which neuromodulation might occur.

Recently, the neuropeptides vasopressin, substance-P, and met-encephalin have been found selectively in the intermediolateral nuclei of the spinal cord, the main outflow tracts of the autonomic system (Anand and Bloom 1984).

Evidence of modulation of nicotinic transmission by non-nicotinic unmyelinated pre-ganglionic fibres was found by Janig, Krauspe, and Wiedersatz (1982). Indeed, these ideas actually predate the discovery of neuropeptides. Costa, Revzin, Kuntzman, Spector, and Brodie suggested in 1961 that a function of noradrenalin in sympathetic ganglia might be to buffer large fluctuations in central sympathetic output. Much more recently, Adams, Brown, and Constanti (1986) have shown a novel potassium channel termed the Im channel. When blocked it leads to a continuous discharge in the transmitting autonomic ganglion cell in response to a maintained depolarization which under normal conditions only produces a short burst of activity. The physiological effect of this channel may therefore be to act as a natural 'brake', limiting the number of repetitive spikes in response to an input. Naturally occurring substances have been found to affect this Im channel.

It is possible that if these physiological buffers to large sympathetic output were to become ineffective, then clinical syndromes like autonomic dysreflexiae would occur, although this approach is likely to be an oversimplification. In addition, Wallin and Stjernberg (1984) suggested that there may be spontaneous fluctuations in excitability in the spinal sympathetic neuron pool. These might underlie the seemingly spontaneous muscle twitches (and spontaneous vasomotor reflexes) observed below spinal injuries and some aspects of autonomic dysreflexiae, and may also be explicable in terms of deranged synaptic control. Differences in end organ responsiveness may also occur and contribute to post-injury syndromes.

Another way of looking at the response of the sympathetic system after spinal cord injury is that there might be differences in tonic (background) and phasic activity. It is known that after injury tonic sympathetic discharge through peripheral nerves to muscle and skin is reduced severely (Wallin and Stjernberg 1984). There may also be an increased responsiveness to certain provocative stimuli, at least in the peripheral vasomotor system, which may underlie dysreflexiae. The lack of a tonic discharge might predispose to a greater response to phasic inputs from the spinal cord, especially if the tonic input was a descending one, from say rostral to the spinal cord, something implied by Bini, Hagbarth, Hynninen, and Wallin (1980) from their microneurographic study. Indeed, they suggested that a function of a central rhythmic autonomic drive under normal circumstances could be to prevent fast repetitive firing of post-ganglionic cells which might be outside the range of maximal responsiveness of end-organs. Phasic and tonic sympathetic ganglion cells have been known for some time. Recently,

Cassell, Clark, and McLachlan (1986) have suggested that the latter group may not have an Im current. Additional evidence for a difference between tonic and phasic cells comes from the work of Dembrovsky, Czachurski, and Sellar (1985), who succeeded in recording from synaptic input cells to autonomic ganglia in the thoracic cord of the cat. They found that there were fewer neurons showing tonic activity in spinal compared to intact animals. It may be relevant in this context that in two patients with complete lesions the introduction of an unphysiological tonic discharge via spinal cord stimulation was found to reduce the incidence of paroxysmal sweating (Illis, Sedgwick, Tallis, and Cole, unpub. obs.).

Thus far, sympathetic cells have been considered almost as a uniform population. However, Janig (1985) in his review was careful to separate the responsiveness of different classes of sympathetic neuron. Most post-ganglionic vasoconstrictor fibres have, for instance, tonic activity, whereas pilomotor and the possible vasodilator ones do not. Reflexes elicited on to the first three cell-types appeared to be intact in spinal animals, but, most interestingly, visceral stimuli which had been inhibitory to peripheral vasoconstriction in intact animals became excitatory in spinal ones. This single seemingly specific alteration in 'local sign' might underlie some of the clinical effects observed after tetraplegia.

In addition to the alterations in synaptic physiology there may be long-term plastic changes whereby neurons produce different axon collaterals once descending control has been removed. Such sprouting was suggested by Johnson, Pallares, Thomason, and Sadove (1975) and quoted in Kewalramani (1980). Sympathetic fibres from the sympathetic ganglion, for instance, have been shown to grow into the hippocampus in rats after lesions in the latter area (Crutcher and Davis 1981). In this case peripheral noradrenergic neurons had replaced central cholinergic ones. One must be careful when making comparisons between species and between different regions of the nervous system. However, this work does suggest that plastic change may occur in the autonomic system after injury. This effect may therefore contribute to some of the alterations observed after spinal injury in man, and would be compatible with the delayed time-scale of some post-injury syndromes.

In conclusion, experimental work suggests potential ways in which autonomic nervous system physiology and hence pathology may alter after spinal cord injury. Dysreflexiae should perhaps not be thought of in terms of a sympathetic system released from descending inhibition to reveal underlying reflex action. There may be changes at synaptic and ganglion levels in tonic and phasic-firing properties, and in the buffering of large fluctuations of sympathetic outflow. Change may also occur in the connections themselves in terms of neuronal plasticity. The recent work from physiological and pharmacological research has been reviewed above to suggest that progress in the understanding of the pathophysiology of the autonomic system after spinal cord injury will best be served by close collaboration between academic as well as clinical researchers and clinicians.

Acknowledgements

I would like to thank Dr L. S. Illis, Dr E. M. Sedgwick, and Dr C. J. Mathias for their constructive criticism of the manuscript, and Prof. D. Whitteridge FRS for his advice and interest.

References

Adams, P. R., Brown, D. A., and Constanti, A. (1982). M-currents and other potassium currents in bullfrog sympathetic neurones. *J. Physiol.* **330**, 537–72.

Amand, P. and Bloom, S. R. (1984). Neuropeptides are selectively markers of spinal cord autonomic pathways. *Trends Neurosci.* **7**, 268–9.

Andersson, P. O., Bloom, S. R., Edwards, A. V., Jarhult, J., and Mellander, S. (1983). Neural vasodilator control in the rectum of the cat and its possible mediation by vasoactive intestinal polypeptide. *J. Physiol.* **344**, 49–67.

Andrews, C. J. H., Andrews, W. H. H., and Orbach, J. (1971). A spinal autonomic reflex evoked by congestion of the mesenteric vein. *J. Physiol.* **213**, 37–8.

Andrews, L. G. and Armitage, K. J. (1971). Sudek's atrophy in traumatic quadriplegia. *Paraplegia* **9**, 159–65.

Anger, R. G., Zehr, J. E., Siekert, R. G., and Segar, W. E. (1970). Position effect of antidiuretic hormone. *Arch. Neurol.* **23**, 513–17.

Beacham, W. S. and Kunze, D. L. (1969). Renal receptors evoking a spinal vasomotor reflex. *J. Physiol.* **201**, 73–85.

Bergman, P., Heilporn, A., Schoutens, A., Paternot, J., and Tricot, A. (1977). Longitudinal study of calcium and bone metabolism in paraplegic patients. *Paraplegia* **15**, 147–59.

Bevan, A. T., Honour, A. H., and Stott, F. H. (1969). Direct arterial recording in unrestricted man. *Clin. Sci.* **36**, 329–44.

Bill, Å. and Linder, J. (1976). Sympathetic control of cerebral blood flow in acute arterial hypertension. *Acta Physiol. Scand.* **96**, 114–21.

Bini, G., Hagbarth, K.-E., Hynninen, P., and Wallin, B. G. (1980). Thermoregulatory and rhythm generating mechanisms governing the sudomotor and vasoconstrictor outflow in human cutaneous nerves. *J. Physiol.* **306**, 537–52.

Bowlby, A. A. (1890). On the condition of the reflexes in cases of injury to the spinal cord. *Med. Chir. Trans.* **73**, 313–25.

Cassell, J. F., Clark, A. L., and McLachlan, F. M. (1986). Characteristics of phasic and tonic sympathetic ganglion cells of the guinea pig. *J. Physiol.* **372**, 457–83.

Chantraine, A., van Ouwenaller, C., Hachen, A. J., and Schinas, P. (1979). Intramedullary pressure and intra-osseous phlebography in paraplegia. *Paraplegia* **17**, 391–7.

Claus-Walker, J. L., Carter, R. E., Lipscomb, H. S., and Vallbona, C. (1968). Analysis of daily rhythms of adrenal function in man with quadriplegia due to spinal cord injury. *Paraplegia* **6**, 195–207.

Claus-Walker, J. L., Carter, R. E., Lipscomb, H. S., and Vallbona, C. (1969). Daily rhythms of electrolytes and aldosterone excretion in men with cervical spinal cord section. *J. Clin. Endocrin.* **29**, 300–1.

Claus-Walker, J. L., Scurry, M., Carter, R. E., and Campos, R. J. (1977). Steady state hormonal secretion in traumatic quadriplegia. *J. Clin. Endocrin.* **44**, 530–5.

Cole, J. D., Mani, R., and Sedgwick, E. M. (1985). Cutaneous vasomotor reflexes following spinal cord injury in man. *J. Physiol.* **369**, 134P.

Connell, A. M., Frankel, H. L., and Guttman, L. (1963). The motility of the pelvic colon following complete lesion of the spinal cord. *Paraplegia* **1**, 98–115.

Corbett, J. L., Frankel, H. L., and Harris, P. J. (1971a). Cardiovascular changes associated with skeletal muscle pain in tetraplegic man. *J. Physiol.* **215**, 381–94.

Corbett, J. L., Frankel, H. L., and Harris, P. J. (1971b). Cardiovascular reflex responses to cutaneous and visceral stimuli in spinal man. *J. Physiol.* **215**, 395–410.

Corbett, J. L., Frankel, H. L., and Harris, P. J. (1971c). Cardiovascular responses to tilting in tetraplegic man. *J. Physiol.* **215**, 411–32.

Corbett, J. L., Frankel, H. L., and Harris, P. J. (1971d). Cardiovascular reflexes in tetraplegia. *Paraplegia* **9**, 113–22.

Corbett, J. L., Debarge, O., Frankel, H. L., and Mathias. C. J. (1975). Cardiovascular responses in tetraplegic man to muscle spasm, bladder percussion and head up tilt. *Clin. Exp. Pharmacol. Physiol.* (Suppl. 2), 189–93.

Corrall, R. J. M., Frier, B. M., McClemont, E. J. W., Taylor, S. J., and Christie, N. E. (1979). Recovery mechanisms from acute hypoglycaemia in complete tetraplegia. *Paraplegia* **17**, 314–18.

Costa, F., Revzin, A. M., Kuntzman, R., Spector, S. and Brodie, B. B. (1961). Role of ganglionic norepinephrine in sympathetic synaptic transmission. *Science* **133**, 1822–3.

Crawford, J. P. and Frankel. H. L., (1971). Abdominal 'visceral' sensation in human tetraplegia. *Paraplegia* **9**, 153–8.

Crutcher, K. A. and Davis, J. N. (1981). Sympathetic noradrenergic sprouting in response to central cholinergic denervation. *Trends Neurosci.* **4**, 70–2.

Cunningham, D. J. C., Guttman, L., Whitteridge, D., and Wyndham, C. H. (1953). Cardiovascular responses to bladder percussion in paraplegic patients. *J. Physiol.* **121**, 581–92.

Dembrovsky, K., Czachurski, J., and Sellar, H. (1985). An intracellular study of the synaptic input to sympathetic preganglionic neurones of the third thoracic segment of the cat. *J. Autonom. Nerv. Syst.* **13**, 201–44.

Dibona, G. B. and Sawin, L. L. (1983). Renal nerves in renal adaption to dietary sodium restriction. *Amer. J. Physiol.* **245**, 322–8.

Dolpuss, P. and Frankel, H. L. (1965). Cardiovascular reflexes in tracheomatized tetraplegics. *Paraplegia* **2**, 227.

Edvinsson, L. (1982). Sympathetic control of cerebral circulation. *Trends Neurosci.* **5**, 425.

Edvinsson, L. and Mackenzie, E. T. (1977). Amine mechanisms in the cerebral circulation. *Pharmacol. Rev.* **28**, 275–348.

Eidelman, B. H. (1973). Cerebral blood flow in normal and abnormal man. DPhil thesis, University of Oxford, Oxford.

Engel, P. and Hildebrandt, G. (1976). Long term studies about orthostatic training after high spinal cord injury. *Paraplegia* **14**, 159–64.

Erulkar, S. D. (1983). The modulation of neurotransmitted release at synaptic junctions. *Rev. Physiol. Biochem. Pharmacol.* **98**, 95–176.

Figoni, S. F. (1984). Cardiovascular and haemodynamic response to tilting and to standing in tetraplegic patients: a review. *Paraplegia* **22**, 99–109.

Frankel, H. L. and Mathias, C. J. (1979). Cardiovascular dysreflexia since Guttman and Whitteridge (1947). *Paraplegia* **17**, 46–51.

Frankel, H. L., Mathias, C. J., and Peart, W. S. (1980). Renin release and sympathetic activity in chronic tetraplegics with cervical spinal cord transection. *J. Physiol.* **308**, 45–6P.

Frankel, H. L., Mathias, C. J., and Spalding, J. H. K. (1975). Mechanisms of reflex cardiac arrest in tetraplegic patients. *Lancet* **ii**, 5183.

Frankel, H. L., Mathias, C. J., Peart, W. S., and Unwin, R. J. (1981). Saralisin induced renal vasoconstriction is due to intrinsic angiotensin-II activity. *J. Physiol.* **319**, 72–3P.

Frier, B. M., Corrall, R. J., Adrian, A. F., Bloom, S. R., Strachan, R. K., and Heading, R. C. (1984). The influence of adrenergic denervation on the response to feeding of the gastroenteropancreatic system in man. *Clin. Endocrinol.* **21**, 639–47.

Gaskell, P. and Burton, A. C. (1953). Local postural vasomotor reflexes arising from the limb veins. *Circ. Res.* **1**, 27–39.

Gilliatt, R. W., Guttman, L., and Whitteridge, D. (1947). Inspiratory vasoconstriction after spinal injuries. *J. Physiol.* **107**, 67–78.

Griffiths, H. J., Bushieff, B., and Zimmermann, R. E. (1976). Investigation of the loss of bone mineral in patients with spinal cord injury. *Paraplegia* **14**, 207–12.

Griffiths, H. J., d'Orsi, C. J., and Zimmermann, R. E. (1972). Use of 125-J photon scanning in the evaluation of bone density in a group of patients with spinal cord injury. *Invest. Radiol.* **7**, 107–17.

Guttman, L. (1963). The paraplegic patient in pregnancy and labour. *Proc. R. Soc. Med.* **56**, 383.

Guttman, L. (1976). *Spinal cord injuries* (2nd edition). Blackwell Scientific Publications, Oxford.

Guttman, L. and Walsh, J. J. (1971). Prostigmin assessment test of fertility in spinal man. *Paraplegia* **9**, 40–51.

Guttman, L. and Whitteridge, D. (1947). Effects of bladder distension on autonomic mechanisms after spinal cord injuries. *Brain* **70**, 361–404.

Guttman, L., Silver, J., and Wyndham, C. H. (1958). Thermoregulation in spinal man. *J. Physiol.* **142**, 406–19.

Harper, A. M., Deshmukh, V. D., Rowan, J. O., and Jennett, W. B. (1972). The influence of sympathetic nervous activity on cerebral blood flow. *Arch. Neurol.* **27**, 1–6.

Head, H. and Riddoch, G. (1917). The autonomic bladder, excessive sweating and some other reflex conditions in gross injuries of the spinal cord. *Brain* **40**, 188–263.

Henrikson, O. (1977). Local sympathetic reflex mechanism in regulation of blood flow in human subcutaneous adipose tissue. *Acta Physiol. Scand.* **S450**, 1–48.

Henrikson, O., Skagen, K., Haxholdt, O., and Dyrberg, V. (1983). Contribution of local blood flow regulation mechanisms to the maintenance of arterial pressure in upright position during epidural blockade. *Acta Physiol. Scand.* **118**, 271–80.

Hermansson, K., Kallskog, O., and Wolgast, M. (1984). Effect of renal nerve stimulation on the activity of the tubuloglomerular feedback mechanism. *Acta Physiol. Scand.* **120**, 381–5.

Hernandez, A. M., Former, J. V., de la Fuenie, T., Gonzalez, C., and Miro, R. (1978). The para-articular ossifications in our paraplegics and tetraplegics: a survey of 704 patients. *Paraplegia* **16**, 272–5.

Hilton, J. (1860). A course of lectures in pain and the therapeutic influence of mechanical and physiological rest in accidents and surgical diseases. *Lancet* ii, 401.

Holmes, C. (1915). Spinal injuries of warfare. *Br. Med. J.* **2**, 769–74, 815–21, 855–61.

Horn, J. P. and Dodd, J. (1983). Inhibitory cholinergic synapses in autonomic ganglia. *Trends Neurosci.* **6**, 180–3.

Hutchinson, J. (1875). Temperature and innulation after crushing of the cervical spinal cord. *Lancet* i, 713–15.

Janig, W. (1985). Organisation of the lumbar sympathetic outflow to skeletal muscle and skin of the cat hindlimb and tail. *Rev. Physiol. Biochem. Pharmacol.* **102**, 119–214.

Janig, W., Krauspe, R., and Wiedersatz, G. (1982). Transmission of impulses from pre- to postganglionic vasoconstrictor and sudomotor neurones. *J. Autonom. Nerv. Syst.* **6**, 95–106.

Johnson, B., Pallares, V., Thomason, R., and Sadove, M. S. (1975). Autonomic hyperreflexia: a review. *Mil. Med.* **140**, 345–9.

Johnson, R. H. (1971). Temperature regulation in tetraplegia. *Paraplegia* **9**, 137–45.

Johnson, R. H., Crampton-Smith, A., and Spalding, J. M. K. (1969). Blood pressure response to standing and Valsalva's manœuvre: independence of the two mechanisms in neurological disease including cervical cord lesions. *Clin. Sci.* **36**, 77–86.

Johnson, R. H., Park, R. M., and Frankel, H. L. (1971). Orthostatic hypotension and the renin–angiotensin system in paraplegia. *Paraplegia* **9**, 146–52.

Jonason, F. (1947). Discussion on the treatment and prognosis of traumatic paraplegia. *Proc. R. Soc. Med.* **40**, 230–1.

Jones, J. V., Fitch, W., Mackenzie, E. T., Strandgaard, S., and Harper, A. M. (1926). Lower limit of cerebral blood flow autoregulator in experimental renovascular hypertension in the baboon. *Circ. Res.* **39**, 355–7.

Kaplan, P. E., Roden, W., Gilbert, E., Richards, L., and Goldschmidt, J. W. (1981). Reduction of hypercalcuria in tetraplegia after weight bearing and strengthening exercises. *Paraplegia* **19**, 289–93.

Kewalramani, L. S. (1979). Neurogenic gastroduodenal ulceration and bleeding associated with spinal cord injuries. *J. Trauma* **19**, 259–65.

Kewalramani, L. S. (198). Autonomic dysreflexia in traumatic myelopathy. *Amer. J. Phys. Med.* **59**, 1–21.

Kramer, H. J., Stinnesbeck, B., Klautke, G., Kipnowski, J., Klingmueller, D., Glaenzer, K., and Duesing, R. (1985). Interaction of renal prostaglandins with renin–angiotension and renal adrenergic nervous system in healthy subjects during dietary changes in sodium intake. *Clin. Sci.* **68**, 387–93.

Krebs, M., Ragnarrson, K. T., and Tuckman, J. (1983). Orthophatic vasomotor responses in spinal man. *Paraplegia* **21**, 72–80.

Langley, J. N. (1893). The arrangement of the sympathetic nervous system based chiefly on observations upon pilomotor nerves. *J. Physiol.* **15**, 176–244.

Lindan, R., Joiner, E., Freehafer, A. A., and Hazel. C. (1980). Incidence and clinical features of autonomic dysreflexia in patients with spinal cord injury. *Paraplegia* **18**, 285–92.

McGarry, J., Woolsey, R. M., and Thompson, C. W. (1982). Autonomic hyperreflexia following passive stretching to the hip joint. *Phys. Med.* **62**, 30–1.

McGregor, J. A. and Meeuwsen, J. (1985). Autonomic hyperreflexia: a mortal danger for spinal cord damaged women in labour. *Amer. J. Obst. Gynecol.* **151**, 330–3.

Mathias, C. J. (1976). Neurological disturbances of the cardiovascular system. DPhil thesis, University of Oxford, Oxford.

Mathias, C. J. and Frankel, H. L. (1983a). Autonomic failure in tetraplegic man. In *Autonomic failure* (ed. R. Bannister), pp. 453–88. Oxford University Press, London.

Mathias, C. J. and Frankel, H. L. (1983b). Clinical manifestations of malfunctioning sympathetic mechanisms in tetraplegia. *J. Autonom. Nerv. Syst.* **7**, 303–12.

Mathias, C. J., Christensen, N. J., Frankel, H. L., and Peart, W. S. (198). Renin release during head up tilt occurs independently of sympathetic nervous activity in tetraplegic man. *Clin. Sci.* **59**, 251–6.

Mathias, C. J., Christensen, N. J., Frankel, H. L., and Spalding, J. M. K. (1979a). Cardiovascular control in recently injured tetraplegics in spinal shock. *Clin. J. Med. (new series)* **48**, 273–87.

Mathias, C. J., Frankel, H. L., Christensen, N. J., and Spalding, J. M. K. (1976). Cardiovascular control in recently injured tetraplegics in spinal shock. *Quart. J. Med.* **48**, 273–87.

Mathias, C. J., Frankel, H. L., Turner, R. C., and Christensen, N. J. (1979b). Physiological responses to insulin hypoglycaemia in spinal man. *Paraplegia* **17**, 319–26.

Mathias, C. J., Christensen, N. J., Corbett, J. L., Frankel, H. L. and Spalding, J. M. K. (1976). Plasma catecholamines during paroxysmal neurogenic hypertension in quadriplegic man. *Circ. Res.* **39**, 204.

Mathias, C. J., Frankel, H. L., Davies, I. B., James, V. H. T., and Peart, W. S. (1981). Renin and aldosterone release during sympathetic stimulation in tetraplegia. *Clin. Sci.* **60**, 399–404.

Mathias, C. J., Christensen, N. J., Corbett, J. L., Frankel, H. L., Goodwin, T. J., and Peart, W. J. (1975). Plasma catecholamines, plasma renin activity and plasma aldosterone in tetraplegic man horizontal and tilted. *Clin. Sci. Mol. Med.* **49**, 291–9.

Meinecke, F. W., Rosenkrantz, K. A., and Kurek, C. M. (1971). Regulation of the cardiovascular system in patients with fresh injuries to the spinal cord—preliminary report. *Paraplegia* **9**, 109–13.

Nanda, R. N., Wyper, P. J., Harper, R. A., and Johnson, R. H. (1974). Cerebral blood flow in paraplegia. *Paraplegia* **12**, 212–18.

Nicholl, R. A. (1982). Neurotransmitters can say more than just 'yes' or 'no'. *Trends Neurosci.* **5**, 369–74.

Niijima, A. and Mei, N. (1984). General involvement of enteroreceptors in motor, homeostatic and behavioural regulations: resumé. *J. Autonom. Nerv. Syst.* **10**, 383–4.

Normell, C. A. (1974). Distribution of impaired cutaneous vasomotor and sudomotor function in paraplegic man. *Scand. J. Lab. Invest.* **33**, S138, 25–41.

Osborn, J. L., Roman, R. J., and Harland, R. W. (1985). Mechanisms of antinatriuresis during low frequency renal nerve stimulation in anaesthetised drugs. *Amer. J. Physiol.* **249**, 360–7.

Pledger, H. G. (1962). Disorders of temperature regulation in acute traumatic paraplegia. *J. Bone Joint Surg.* **44B**, 110–13.

Pollock, L. J., Boshes, B., Chor, H., Finkelman, I., Arieff, A. J., and Brown, M. (1951). Defects in regulatory mechanisms of autonomic function in injuries to the spinal cord intubations. *J. Neurophysiol.* **14**, 85–93.

Prigioni, I. and Casella, C. (1984). Acetylcholine-induced back firing in the pre-ganglionic trunk of the cat superior cervical ganglion. *J. Autonom. Nerv. Syst.* **10**, 19–25.

Riddoch, G. (1917). The reflex function of the completely divided spinal cord in man, compared with those associated with less severe lesions. *Brain* **40**, 264–402.

Rowe, M., Robinson, J. E., Ells, P., and Cole, J. D. Pain following spinal cord injury; results from a postal survey. *Paraplegia* (in press).

Rowell, L. B. (1984). Reflex control of regional circulations in humans. *J. Autonom. Nerv. Syst.* **11**, 101.

Scher, A. T. (1976). The incidence of ectopic bone formation in post traumatic paraplegic patients of different racial groups. *Paraplegia* **14**, 202–6.

Sharpey-Schafer, E. P. (1961). Venous tone. *Br. Med. J.* **11**, 1589–95.

Silver, J. R. (1970). Vascular reflexes in spinal shock. *Paraplegia* **8**, 231–42.

Skagen, K., Jensen, K., Henrikson, O., and Knudsen, K. (1982). Sympathetic reflex control of subcutaneous blood flow in tetraplegic man during postural changes. *Clin. Sci.* **62**, 605–9.

Stella, A., Golin, R., Busnardo, J., and Zanchetti, A. (1984). Effects of afferent renal nerve stimulation on renal hemodynamic and excretory functions. *Amer. J. Physiol.* **247**, 576–83.

Uprus, V., Gaylor, J. B., Williams, D. J., and Carmichael, E. A. (1935). Vasodilatation and vasoconstriction response to warming and cooling the body. A study in patients with hemiplegia. *Brain* **58**, 448–69.

Verkuyl, A. (1976). Sexual function in paraplegia and tetraplegia. In *Handbook of clinical neurology* (ed. P. J. Vinken and G. W. Brugn), pp. 437–62. North-Holland, Amsterdam.

Wallin, B. G. and Nerhed, C. (1982). Relationship between spontaneous variations of muscle sympathetic activity and succeeding changes of blood pressure in man. *J. Autonom. Nerv. Syst.* **6**, 293–302.

Wallin, B. G. and Stjernberg, L. (1984). Sympathetic activity in man after spinal cord injury. *Brain* **107**, 183–98.

Wallin, B. G. and Sundlof, G. (1982). Sympathetic outflow to muscles during vasovagal syncope. *J. Autonom. Nerv. Syst.* **6**, 287–92.

Walsh, J. J. (1960). Cardiovascular complications in paraplegia. Proceedings of Scientific Meeting, International Stoke Mandeville Games, Rome, pp. 37–45.

Watson, N. (1981). Late ileus in paraplegia. *Paraplegia* **19**, 13–16.

Wayne, W. M. and Vukor, J. G. (1977). Eye findings in autonomic hyperreflexia. *Ann. Ophthalmol.* **9**, 41–2.

Welply, N. C., Mathias, C. J., and Frankel, H. L. (1975). Circulatory reflexes in tetraplegics during artificial ventilation and general anaesthesia. *Paraplegia* **13**, 172–82.

Urodynamic and neurophysiological assessment of the neuropathic bladder

J. A. Massey

Introduction

In the latter half of this century the life-expectancy of patients who have suffered spinal cord injury has increased dramatically. Hackler (1977) reported a 63 per cent mortality at 25 years for 170 Second World War paraplegics, but only 26 per cent for 100 Korean War patients at 20 years. Improvements in medical care, directed especially towards the prevention and treatment of respiratory problems and pressure sores, have contributed greatly, but a major advance has been in our understanding and, therefore, management of the lower urinary tract dysfunction which leads to renal failure. Preservation of renal function must be the prime objective of all urological management.

The improved life-span of these patients has led to greater concern for their quality of existence. Although it must always remain a secondary consideration, wherever possible it should be our aim to achieve a dry patient, as independent of help from others as possible and devoid of an appliance, objectives which not only improve self-image and morale, but may themselves aid in the prevention of ulceration and infection.

Lower urinary tract function

Management of the compromised urinary tract should be designed to emulate as nearly as possible its normal role. In essence the vesicourethral unit provides a low-pressure reservoir of 'good' capacity (approximately 500 ml with a pressure increment of less than 10 cm H_2O). Voiding occurs at will and to completion, thereby providing an adequate functional capacity and obviating the opportunity for infection in a residual. It occurs without an undue rise in detrusor pressure because of synergic relaxation of the sphincteric mechanisms. High pressures with or without dyssynergic sphincter activity lead to vesicoureteric reflux and renal damage. Both anatomical and neurological factors influence this function (Fig. 10.1).

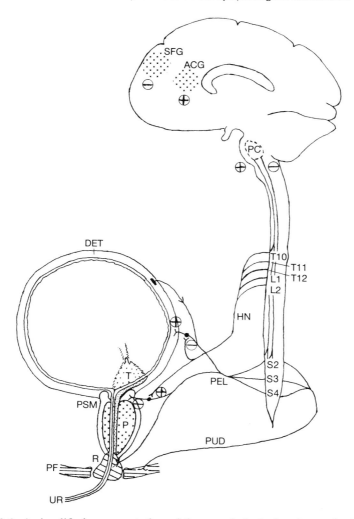

Fig. 10.1. A simplified representation of the morphological and neurological components of micturition. Key: ⊕, facilitation; ⊖, inhibition; ACG, anterior cingulate gyrus; DET, detrusor muscle; HN, hypogastric nerve; P, prostate; PC, pontine centres; PEL, pelvic nerve; PF, pelvic floor; PSM, proximal sphincteric mechanism; PUD, pudendal nerve; R, rhabdosphincter; SFG, superior frontal gyrus; T, trigonal smooth muscle.

The detrusor

The whole of the dome of the bladder, including the muscle deep to the trigone, consists of a meshwork of smooth muscle cells with no distinct layers and which contract *en masse* (Gosling, Dixon, and Humpherson 1983). The detrusor is richly supplied by cholinergic (parasympathetic) excitatory nerves with their cell bodies in both the pelvic autonomic plexus and the bladder wall (Taira 1972). Pre-ganglionic fibres run in sacral roots 2–4. Sympathetic noradrenergic nerves are sparse in the bladder vault and accompany the blood vessels (Sundin, Dahlstrom, Norlen, and Svedmyr 1977). They arise at the T10–L2 level, running in the pre-sacral nerves and hypogastric plexus. There appears to be no direct action on the detrusor, but they may have an inhibitory influence in the pelvic ganglia. There is some evidence for non-adrenergic, non-cholinergic, nerve-mediated effects which may indicate a third type of effector nerve or an alternative transmitter in 'cholinergic' nerves. The afferent nerve supply is ill-understood (Gosling *et al.* 1983) but virtually all conscious sensation in the human enters the S2–4 roots via the pelvic nerves (Iggo 1955). However, the feeling of imminent voiding (urgency) appears to originate in the urethra or periurethral striated muscles with afferent fibres running in the pudendal nerves and dorsal columns of the spinal cord (Nathan 1956).

The proximal sphincteric mechanism

The superficial trigonal muscle has fewer cholinergic fibres than the detrusor and relatively frequent noradrenergic nerves. This 'reversal' of innervation is more marked in the male than the female bladder neck region. In the female the smooth muscle fibres are aligned longitudinally. In the male they encircle the bladder neck and are continuous with the prostatic capsule, consistent with their role as a 'genital sphincter' preventing reflux of ejaculate into the bladder. Despite experimental studies with alpha-blockers (Donker, Ivanovici, and Noach 1972), suggesting that up to 80 per cent of resting urethral pressure in both sexes depends on sympathetic activity, the clinical use of alpha-blockade in neuropathic bladders has been generally disappointing.

The distal sphincteric mechanism

This mechanism comprises two groups of striated muscle. Surrounding the urethra but separated from it by connective tissue is the periurethral striated muscle, part of the pelvic floor. The pudendal nerve (S2–4) supplies a mixture of 'fast-twitch' and 'slow-twitch' muscle fibres with muscle spindles. It is capable of rapid activity to interrupt the stream.

The intrinsic distal sphincter or rhabdosphincter, on the other hand, has only small diameter, 'slow-twitch' fibres without muscle spindles and is able to maintain tone over longer periods. It is supplied from the same cord segments but via the pelvic nerves. In the male the muscle fibres form a circular sphincter distal to the prostate and interdigitate with muscle from the anterior and anterolateral aspects of the prostate. There is no such annular arrangement in the female, the muscle bulk being deficient posteriorly. The greatest condensation is in the middle third of the urethra.

The prostate and urethra

Urethral closure is not dependent solely on the sphincter mechanisms outlined above but also involves passive factors such as urethral softness and elasticity, and mural tension. In the female the urethra is a relatively short straight tube so that obstruction is uncommon but incontinence is frequent. Lack of oestrogenization may compound the problem by its effect on the mucosa. In the male its greater length and curves may contribute to inefficient emptying. Older male patients may have or acquire an obstructive element due to prostatic hypertrophy, and previous catheterization may induce strictures, in addition to any problem caused by the underlying neuropathy.

Central nervous control

Blaivas (1982), in clinical and urodynamic studies in 550 patients, evinces evidence that normal micturition is a brain-stem reflex under voluntary control. Experimental studies in animals and the results of various lesions in humans have identified numerous areas in the brain responsible for facilitation and inhibition of vesicourethral function.

Conscious awareness of micturition status and the meeting of social mores involve the superior frontal gyrus and anterior cingulate gyrus. At a subcortical level several areas from the septal region through to the pontine reticular formation and cerebellum co-ordinate bladder function. The concept of a pontine micturition centre is thus oversimplistic, but its disconnection from the cortex produces initial areflexia followed by hyperreflexia with co-ordinated sphincter activity. These terms are defined in the section on urodynamic studies.

The brain-stem centres are capable of independent 'automatic' activity, co-ordinating and controlling detrusor and sphincter activity. Spinal cord injury above the sacral region produces the areflexia of spinal shock followed by detrusor hyperreflexia. The strength and duration of such bladder contractions appears to depend on the number of intact segments.

The sphincters normally relax just prior to, and throughout, a voluntary voiding contraction. The sacral micturition centre (Denny-Brown and Robertson 1933) situated in the *conus medullaris* is capable of completely autonomous reflex function but cannot fully co-ordinate parasympathetic, sympathetic, and somatic activity. If the nerves concerned with the distal sphincteric mechanism are involved then a failure of co-ordination with detrusor contraction (detrusor/sphincter dyssynergia) occurs. Detrusor/ bladder neck dyssynergia is uncommon in cord-injured patients.

Disruption of the nervous pathways below the sacral centre removes both the bladder and sphincters from neurological 'drive'. The result is a non-contracting bladder and a sphincter that neither contracts nor relaxes.

Contributory factors

In addition to these neurourological considerations, dysfunction may be exacerbated by other factors in the spinal cord injured patient. High lesions may impair the contribution of abdominal muscle and diaphragmatic contractions to the expulsive effort. On the other hand, the weight gain that is commonly seen may render a compromised urethra incompetent under stress. Incontinence may also occur in conjunction with infection and stones. These problems are often magnified by a failure to take into account the immobility and mental capacity of the patient. The intellectual and/or physical dexterity required to understand and manipulate some of the devices and appliances available, and the need for cleanliness and a reasonable fluid intake, all require individual assessment.

The urological *sequelae* of spinal cord injury

Following the period of spinal shock with absent bladder function when most patients will require a catheter, there is a period of recovery. Catheterization may be intermittent or, less commonly, in-dwelling per urethram or suprapubically. The bladder over a variable period of time achieves a balanced state which varies with the involved site.

A complete suprasacral lesion will result in a senseless bladder with its capacity limited by spontaneous detrusor contractions. Compliance is usually satisfactory. Infection and cold contribute to a lowered capacity. With high lesions (quadriplegia), brain-stem co-ordination may allow synergic detrusor sphincter activity, but at lower levels voiding efficiency is compromised by detrusor/sphincter dyssynergia. The high-pressure system that ensues may be sufficient to cause vesicouretric reflux and renal impairment even in the absence of infection. In many patients their detrusor contraction is not sustained, leading to residual urine. Sphincter contraction may itself be leading to detrusor inhibition (McGuire and Morrissey 1982) rather than the lack of higher centre input.

A complete lower motor neuron lesion again produces an insensate bladder, but of high capacity and acontractile (areflexic). The sphincter is inactive and voiding can occur by strain. The low pressures generated in such a system, although predisposing to residuals and infection, do not commonly pose a problem with reflux. However, two factors may complicate the situation. Compliance may be poor, although this is commonest in incomplete *cauda equina* lesions, and the urethra is not always completely 'relaxed' so that a dual pathology exists with stress incontinence and a degree of obstruction. Such a situation has been termed 'isolated distal sphincter obstruction'. Postulated causes of this include denervation fibrosis and sympathetic overactivity. The role of urethral smooth muscle is poorly understood, but evidence that it may be contributory is accumulating (Awad and Downie, 1977; Sunder, Parsons, and Gibbon 1978; Koyanagi, Arikado, Takamatsu, and Tsuji 1982).

Complete cord lesions are, however, relatively uncommon and this is reflected in the range of bladder and urethral activity witnessed during urodynamic assessment of these patients. In terms of management, the functional description these studies allow is of far greater significance than the historical or neurological localization of the level of injury. This does not mean that the latter should be ignored but rather interpreted with caution. All studies should be preceded by a careful urological history and include a detailed examination of the patient, the most important features of which are the presence or absence of a palpable bladder and whether the ano-cutaneous reflex is present or absent. A positive response to pinprick at the anus (reflex contraction) is indicative of a suprasacral lesion and usually denotes detrusor hyperreflexia. Conversely, a negative test portends detrusor areflexia. An estimation of physical and psychological status and intelligence at this stage will aid the determination of suitable treatment modalities indicated by the tests. All patients will require intravenous urography to assess the upper tracts and urine culture.

Urodynamic studies

Physiological investigation of the lower urinary tract is interpreted in terms of capacity, sensation, compliance, and contractility of the detrusor, efficiency of voiding (detrusor contraction and urethral relaxation), and secondary effects such as reflux.

The filling phase

Figure 10.2(a) illustrates a filling cystometrogram of normal compliance and capacity. In Fig. 10.2(b) capacity is limited by involuntary detrusor contractions, culminating in leakage, which in the presence of neurological disease is termed detrusor hyperreflexia. Poor compliance alone, Fig. 10.2(d), is rarely a problem in neuropathy and frequently, if present, is artefactual due to rapid filling [Fig. 10.2(c)]. Fig. 10.2(e) illustrates an areflexic bladder (large capacity, no detrusor contraction).

The total amount of fluid the bladder will hold under test conditions is termed the maximum cystometric capacity (MCC). The functional capacity is considerably less than this if residual urine is present after voiding (functional capacity = MCC − residual).

Sensation is not normal in these subjects. Abdominal fullness may be the indicator for voiding or the 'urgency' occasioned by hyperreflexic

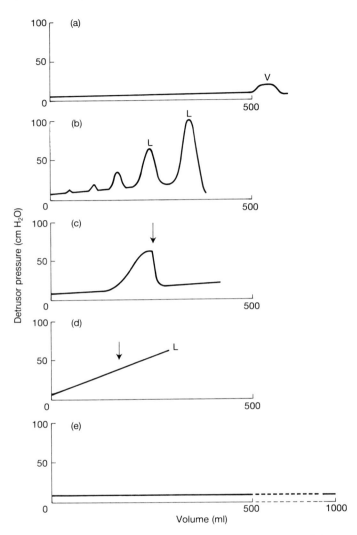

Fig. 10.2. Diagrammatic representation of cystometric patterns. (a) Normal with low-pressure voluntary void at V. (b) Detrusor hyperreflexia with involuntary leakage at L. (c) Artefactual compliance change revealed by stopping fill at arrow. (d) Hypocompliance (no change by stopping fill at arrow) with leakage, L, when vesical pressure exceeds sphincter pressure. (e) Hypercompliant (areflexic) bladder.

contractions, but neither of these are equivalent to the normal person's first or final desire to void. As previously stated, there may be no awareness even of high-pressure detrusor contractions. Some individuals develop a bizarre signal such as warmth on the top of the head or a somatic muscle spasm.

The compliance is defined as volume increment per unit pressure rise (ml/cm H_2O). The contribution of the active and passive components of the bladder wall to its compliance is an area of great interest at present, which is to say that little is understood of the phenomenon. Fibrosis of the detrusor from any cause will produce the appearance illustrated in Fig. 10.2(d) whatever the speed that fluid enters the bladder, whereas the filling rate more commonly is related to the degree of hypocompliance [Fig. 10.2(c)]. Stopping the fill and proceeding at a slower rate will be indicative of this situation and should be undertaken whenever low compliance changes are noted. The adrenergic system could be implicated in these cases, for Sands, Constantineu, and Govan (1972) showed that the (sympathetic) haemodynamic changes that occur with vesical distension are sensitive to its rate rather than to its absolute volume.

Detrusor contraction is normally absent during the filling/storage phase of bladder function. Numerous attempts to quantify abnormal contractions have been made (Abrams 1977), but a universally acceptable 'instability index' has not been agreed. Some attempt should be made to annotate the detrusor pressures reached and their duration, and whether leakage occurs. For these purposes it is essential to measure intravesical (P_{ves}) and intra-abdominal (P_{abd}) pressures, the latter usually approximated by a rectal pressure line. Electronic subtraction allows the detrusor pressure alone (P_{det}) to be distinguished from movement or other artefacts.

If leakage does occur, the above measurements enable its attribution to hyperreflexic (detrusor) or abdominal pressure rises. In the latter case urethral (stress) incompetence is present. With care it is possible to measure bladder and urethral (P_{ura}) pressures synchronously (urethrocystometry). In practice little additional information is obtained, especially if radiographic visualization is employed with pressure studies.

The voiding phase

The flow rate (Q) is expressed as volume of urine voided (ml) per second (s). The peak achieved (Q_{max}) and the average flow rate (Q_{ave}) may be noted. They are most reliable in the range 200–400 ml and the absolute values are also dependent on age and sex. Urinary flow occurs as the summation of the expulsive forces (abdominal pressure, detrusor contraction, and duration) and urethral relaxation. Normal flow is continuous but abnormalities of either half of the above equation may result in intermittent flows. Voiding time is defined as the total duration of micturition. Hence, in this situation, flow time may be considerably less than voiding time. The presence of residual urine after voiding should be noted.

Urodynamic techniques

The exact methods employed to obtain the above information differ in detail from centre to centre but are uniformly designed to be non-provocative; that is, they attempt to reproduce the patient's normal storage and voiding cycle so far as possible. Whilst continuous monitoring (James 1979) would have attractions for this purpose, the additional information that can be obtained with synchronous radiological visualization or electromyography of the sphincteric region outweighs the disadvantages of limiting the investigation in time and space. This is counteracted by repeated fills until a stable voiding pattern is obtained.

Pre-study considerations

The necessity for regular urine culture has already been mentioned. Patients with active infection should be treated effectively prior to the investigation, both as a precaution against bacteraemia and to eliminate its effect on detrusor contractility and compliance. An asymptomatic bacteriuria is best 'covered' by a single dose of appropriate antibiotic about one hour before the procedure.

Patients with suprasacral cord lesions are able to achieve control over their large bowel function by a high-fibre diet and the use of suppositories to precipitate elimination. The timing of the investigation should allow for the fact that flowmeters are designed specifically for urine. Similarly patients with lower level injuries managed on a low-residue diet may evacuate by strain or by manual clearance. This is best done before the test.

The position in which the patient is investigated is dictated by their disability/deformity; the facility with which they can stand, sit unsupported, or need to lie supine. It may be impossible to perform a seated test satisfactorily in a patient with severe kyphoscoliosis. Patients with marked abdominal or limb spasms are best investigated supine and will often require suprapubic rather than urethral access to the bladder. This should be achieved at least forty-eight hours beforehand.

Flowmetry

The use of uroflowmetry alone or with ultrasound measurement of residuals is inadequate for diagnosis but is an easily repeated non-invasive measure for following the progress of a treatment modality. Wherever possible it should be used in conjunction with synchronous pressure studies. In the recumbent position we achieve this by the use of a length of wide-bore plastic pipe between the penis and flowmeter. Any of the available types (spinning disc, weight transducer, or capacitance; Massey and Abrams

1985) are suitable, but interpretation of the hard-copy readout must allow for the time taken for urine to traverse this conduit. There is no satisfactory method of achieving a measured flow in the recumbent female.

Urethrometry

Static urethral pressure profile measurements provide a picture of the total activity of the urethra. We commonly employ the perfusion technique of Brown and Wickham (1969), but they may equally well be determined by catheter-mounted microtransducers, withdrawn at 1 mm/s in the female or at 3 mm/s in the male. Rossier and Ott (1976) have modified the perfusion method for synchronous recording of bladder and urethral pressures in spinal cord injuries, whilst Bary, Day, Lewis, Chawla, Evans, and Stephenson (1982) have used the dual microtip transducer catheter. The latter showed a sudden increase in urethral pressure prior to detrusor contraction in the majority of patients with reflex detrusor activity, which was inferred to arise in the periurethral striated sphincter since sphincterotomy did not abolish it. It is suggested that the sphincter contraction may be the trigger to the detrusor contraction in these patients. Such facts must, however, be interpreted with caution for the catheters are directional and changes in orientation as well as position can significantly alter the recorded pressure. The devices, too, are relatively stiff and may introduce artefacts by their 'straightening' effect on the urethra or by their own bending. Schafer, Gerlach, and Hannapel (1982) have shown that they are inaccurate for the measurement of intravesical pressure because of these factors and the unknown position of the tip transducer. The indications for all these studies in isolation are again limited and it may even be impossible to negotiate the distal sphincter spasm of some patients with suprasacral lesions. Their value lies in the assessment of (i) the result of surgical ablation of the distal sphincter, (ii) the efficacy of alpha-adrenergic blockade in patients with absent *conus* activity, and (iii) the localization of problems persisting after insertion of an artificial sphincter.

Cystometry

Access may be per urethram or suprapubic, using small-bore fluid-filled tubes connected to external pressure transducers or employing microtip catheter transducers. A rectal line is inserted and taped into position to approximate intra-abdominal pressure. The recording equipment has the facility for electronic subtraction to provide detrusor pressure recording on the hard-copy (paper trace) computer printout or video recording.

A filling line is also required. It is our practice to use the 8 Ch. balloonless catheter employed for the profile studies. The 1.2 mm external diameter pressure line may be piggy-backed into the bladder with this. Thomas (1979) recommends using two identical catheters of the same diameter or a 6 Ch.

catheter alone if this proves impossible. With the latter method it is necessary to stop filling intermittently to eliminate artefacts caused by the instillation. The same author has stressed the finely balanced vesicourethral function in the neuropathic bladder. Several modifications of technique from the normal provocative studies are, therefore, recommended. The residual urine should not be drained prior to filling but may be estimated ultrasonically. Bladder emptying may alter the pattern of reflex detrusor activity and detrusor/sphincter dyssynergia. The only exceptions to this rule are where the utilization of intermittent self-catheterization or insertion of an artificial sphincter are contemplated. The effect of filling rate on compliance has already been mentioned, but rapid distension may also provoke hyperreflexia and dyssynergia. A slow fill of 10 ml/min is employed in adults and frequently lower than this in children. Control of filling rate is best achieved using a peristaltic pump. The introduced volume may then be accurately known at a given juncture by timing a calibrated pump if transducer mounting for the instillation fluid is not available. The initial fill and void study after insertion of the catheter may not be representative. Hence several sequences are repeated with the catheters *in situ* to establish a reproducible voiding pattern.

Videocystometry
The addition of radiological surveillance to standard cystometry, linked to a video recorder, enormously enhances the value of either alone and has become standard practice in the investigation of lower urinary tract function in the spinal cord injured patient. Indeed the information supplied by this combination is so good that we rarely use sphincter electromyography. A 30 per cent contrast solution (urografin 150) has been found to provide adequate contrast, even in the presence of a large residual urine, using a tilting table or C-arm image intensifier. An oblique position enables clear visualization of the bladder neck and urethra. The effect of alteration of position may also be observed, which in some patients can induce significant changes in vesicourethral function. Information is obtained on bladder size and outline, the presence of bladder neck or sphincter incompetence, the site and degree of outflow obstruction, and the presence of intraprostatic and vesicoureteric reflux.

The two methods by which it is achieved in the Clinical Investigation Unit in Bristol are illustrated in Fig. 10.3.

Electromyography (EMG)
The bioelectric potentials generated by depolarization of striated muscle may be measured by either surface or needle electrodes with a further distant ground electrode. The former may be mounted on an anal plug, on a urethral catheter, or on adhesive pads placed on the perineum. They

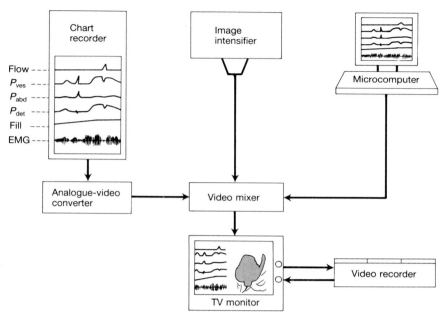

Fig. 10.3. The two methods by which synchronous video urodynamic studies are performed in the Clinical Investigation Unit, Ham Green Hospital, Bristol.

measure mass activity in the underlying muscle and as such may not be representative of urethral sphincter activity (Vereecken and Verduyn 1970), whilst movement of catheter electrodes may produce artefacts or stimulate sphincter contraction. Acrylic anal plug electrodes can modify lower urinary tract function and elicit reflex anal sphincter contraction and dislodgement of the plug. Needle electrodes may be concentric or single fibre and sample a small number of motor units. The external anal sphincter, *levator ani*, or urethral rhabdosphincter have been studied, although the latter is difficult to locate accurately and to maintain a constant position within. The interpretation of these studies is directly proportional to experience, since artefacts occur from injury potentials due to local trauma and the pain of needle insertion. They are open to the criticism that the sample may not be representative of the muscle bulk.

Voluntary voiding is normally preceded by complete electrical silence, following a period of gradually increasing EMG activity with filling. Coughing, voluntary sphincter contraction, and a number of other manœuvres increase EMG activity. There is a complete absence of electrical activity immediately following a lower motor neuron lesion. Thereafter, a period of spontaneous activity (sharp waves or fibrillation potentials on needle EMG) follows. Reinnervation is signified by polyphasic potentials in incomplete lesions and the return of reflex activity is only partial. In suprasacral lesions there is no voluntary sphincter control. EMG studies are of greatest value in revealing (by exclusion) a smooth muscle component to distal urethral obstruction and in the assessment of detrusor/sphincter dyssynergia, the involuntary contraction of the distal sphincter mechanism during an hyper-reflexic contraction. Blaivas, Sinha, Sayed, and Labib (1981) have described three types of such activity (DSD). In type-I DSD, EMG activity peaks at the height of the detrusor contraction, then completely relaxes allowing voiding as detrusor pressure falls. Sporadic bursts of activity occur in type-II DSD throughout the detrusor contraction, whilst in type-III DSD the EMG activity waxes and wanes in parallel with detrusor waves, so that often no void occurs.

Either type of electrode will require a specific filtered amplifier. In conjunction with videourodynamic studies surface electrodes producing an integrated trace provide an adequate quantitative readout. Qualitative judgements on motor units require an oscilloscope and high-frequency recording equipment.

Electromyelography

In 1972 Bradley introduced this technique to measure the integrity of the sacral reflex arc. In the UK they are more commonly known as sacral-evoked responses (SERs). Although primarily designed to diagnose the subtle neuropathies that may occur, perhaps in isolation, in such conditions as diabetes mellitus and multiple sclerosis, SERs have been used in spinal cord injured patients. It retains largely an experimental role in such cases. The original technique involved stimulation by catheter-mounted electrodes at 4–15 V with pulses of 100–200 msec duration. The EMG response of the anal sphincter was ascertained by an anal plug electrode. Many other sites of peripheral stimulation using block or needle electrodes have been described. Similarly a multiplicity of responses have been recorded, usually a muscle contraction, but Eisen and Elleker (1980) have also measured spinal cord and cerebral cortical-evoked potentials using surface electrodes. The various potential combinations are listed in Table 10.1. Each of these combinations has a specific conduction time or latency between the peripheral stimulation and the recorded response dependent on the number of synapses involved, although some are less stable than others. The intensity of stimulation is gradually increased and sensory threshold, reflex threshold, and the minimum latency are recorded for a series of studies.

In complete lower motor neuron lesions evoked responses are absent. If the lesion is incomplete latency and sensory thresholds may be normal or increased and it may be possible to demonstrate asymmetric involvement.

Table 10.1. Potential sites for evoked response studies

Stimulation site	Response site
Glans penis	Bulbocavernosus muscle
Dorsal nerve of penis/clitoris	Anal sphincter
Urethral *mucosa*	Distal urethral sphincter
Bladder *mucosa*	*Levator ani*
Cutaneous leg nerves	Cerebral cortex
	Spinal cord

Upper motor neuron lesions commonly exhibit diminution of the latency, presumably because fewer synapses are involved (Krane and Siroky 1980).

Additional tests

These are mentioned for the sake of completeness but find their commonest application in non-traumatic known or covert neuropathy.

The *ice water test* has been used in cord injuries. Instillation of very cold fluid into the bladder provokes hyperreflexia.

Denervation supersensitivity tests have been described for both the detrusor (Lapides, Friend, Ajemian, and Reus 1962) and urethra (Koyanagi 1978). Both require a standardized procedure and some experience of the technique for reliability. When smooth muscle is denervated its sensitivity to neurotransmitters, drugs, and ions is increased, the more so when this is post-ganglionic rather than pre-ganglionic (Koizumi and Brooks 1974). A method is thereby available for distinguishing between neuropathic and myopathic underactivity. In testing the detrusor, 0.25 mg carbachol is given subcutaneously, with a standardized volume in the bladder, and is positive if a detrusor pressure rise greater than $20\,cm\,H_2O$ is induced over thirty minutes. It cannot be used in asthmatics, cardiac disease, or hyperthyroidism. For the urethra, 4 mg of ethylphenylephrine or an equivalent dose of an alpha-adrenergic drug is administered. A positive response generates an increase of $15\,cm\,H_2O$ or more in maximum urethral closure pressure over five minutes.

The *phentolamine test* may assist in determining whether smooth or striated muscle is contributing dominantly to relative outlet obstruction by the effect of alpha-adrenergic blockade on urethral relaxation (Olsson, Siroky, and Krane 1977).

Anticholinergic tests do not correlate well with clinical practice or distinguish between detrusor hyperreflexia and idiopathic detrusor instability. Intravenous propantheline will abolish contractions in both.

The role of urodynamic studies in practice

It will be evident from the foregoing that urological symptoms in cord injury patients may be confusing, especially in the presence of sensory loss. Incontinence and retention of varying degrees may co-exist and upper tract damage may develop insidiously, particularly in the male patient, commonly with no observed change in voiding habit.

The functional assessment of storage capacity and pressure, voiding function and pressure, and sphincteric activity that these studies allow are essential in the logical management and surveillance of these patients. The practical aspects of treating spinal cord damaged outflow tract neuropathy are best considered under the two main types dependent on the level of the lesion.

Suprasacral cord lesions

Inability to void This should be attributable to the absence of detrusor contractility or severe degrees of detrusor–sphincter dyssynergia. The former may exist simply because the 'recovery stage' is not sufficiently progressed and the decision must be made whether to persist with self-intermittent catheterization (SIC) or to make attempts to induce vesical contraction with drugs or by 'trigger' mechanisms. The latter vary from patient to patient and may include suprapubic tapping, stroking the glans or thigh, tugging the pubic hair, anal dilatation, and sniffing. Severe DSD will require external sphincterotomy.

Inefficient reflex voiding This results in residual urine and the dangers of infection. There is no consensus on what constitutes a significant residual. Personal experience may suggest an absolute value or reliance on symptoms and upper tract surveillance. A weak and poorly sustained reflex contraction may be enhanced by drugs or electrical (sacral anterior root) stimulation. Their efficacy must be monitored urodynamically and the latter will require the studies perioperatively (to identify the appropriate sacral roots) and post-operatively to set stimulation levels and to check continued function of the device. Emptying limited by functional outlet obstruction can be localized to the distal sphincter (DSD—increased EMG activity—Fig. 10.4), bladder neck area, or urethra (silent voiding EMG). The latter are rare in spinal cord injury but may be responsive to alpha-adrenergic blockade. Videourodynamics define whether surgery should be directed to the sphincter or bladder neck. Intra-operative profilometry may be used to check the completeness of the procedure or alternatively a repeat video study will give information on its functional result. On occasion there may be a delayed return of distal sphincter activity, which may be due to an inadequate initial incision in its length or depth, or 'reactivation' of striated

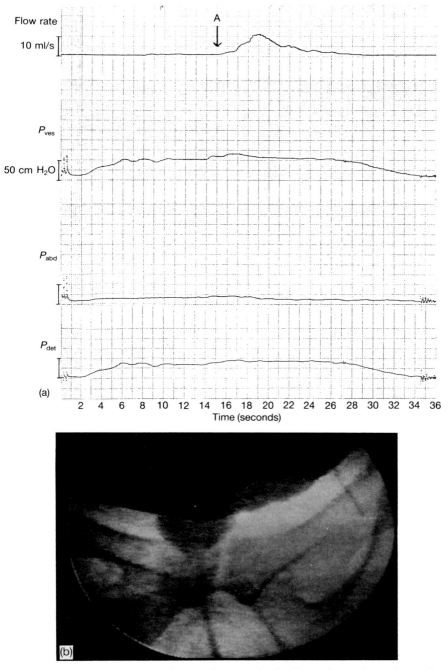

Fig. 10.4. Detrusor/external sphincter dyssynergia. Radiograph at point A shows the typical 'spinning-top' appearance of the proximal urethra. Hyperreflexic detrusor contraction low but sustained. Tetraplegia C4. (a) Cystometrogram; (b) still from cineoradiograph.

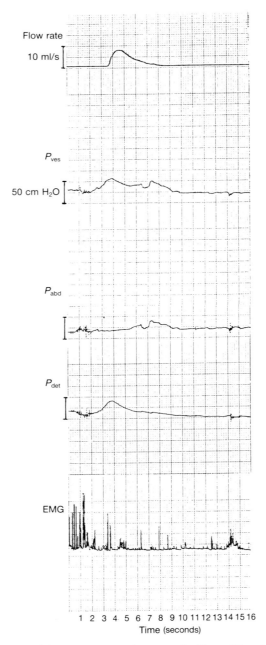

Fig. 10.5. Unsustained detrusor contraction provoked by sniffing. Note 'silent' EMG activity during void of 70 ml at a rate of 8 ml/s. Residual urine, 200 ml.

muscle activity. Sometimes adequate surgery may fail because of poorly sustained detrusor contractions (Fig. 10.5). The level of reflex activity and dyssynergia often changes with time, most usually presenting with the onset of recurrent infections. This or a changing voiding pattern require reassessment by urodynamics.

High-pressure voiding This is perhaps the most dangerous situation. The patient may remain continent and void to completion but at the expense of generating very high voiding pressures in the presence of DSD (Figs 10.6 and 10.7). The pressures generated may be sufficient to engender upper urinary tract dilatation and deterioration, even in the absence of reflux. In some centres a policy of careful follow-up is employed, but it is the practice of most specialists to undertake early sphincterotomy, for antispasticity drugs have been accorded little success. It is often a difficult decision in male patients with partial lesions who have retained sufficient sensation to remain continent. The latter is usually dependent on their dyssynergia, and sphincter ablation often renders them incontinent and necessitates a condom appliance. In these patients, particularly, regular check-ups on renal function with renal isotope studies, intravenous urography, and videourodynamics are mandatory.

Autonomic dysreflexia This may occur in patients with high dorsal or cervical lesions. A stimulus below the level of the lesion, usually rectal or vesical, produces an exaggerated sympathetic response with hypertension, bradycardia, headache, and profuse perspiration. The syndrome has been known to be fatal due to cerebral haemorrhage. The prevention of retention is essential and urodynamics may relate more minor degrees of the syndrome to hyperreflexic contractions, and particularly dyssynergia. Transurethral sphincterotomy renders considerable relief in such instances.

Female incontinence This is a thorny problem due to the inadequacy of urine collecting appliances. If reflex voiding can be shown to occur at a specific and reasonable volume, then triggered voiding or SIC can be employed 'by the clock'. Anticholinergics are effective in some patients in increasing vesical capacity and reducing hyperreflexic incontinence. It may be preferable to render the detrusor underactive by this means or by denervation procedures (sacral nerve blocks, sacral neurectomy, or intrathecal blocks) and employ straining, the Crede manœuvre, or SIC for emptying the bladder.

Lower motor neuron lesions
Although SIC is the treatment of choice in these patients, their aesthetics, manipulative abilities, or constitution may mitigate against its employment.

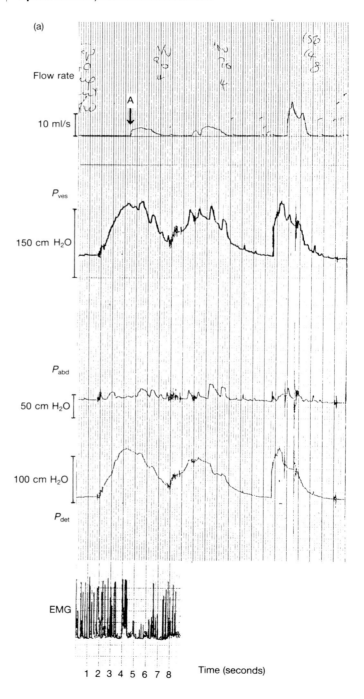

(a)

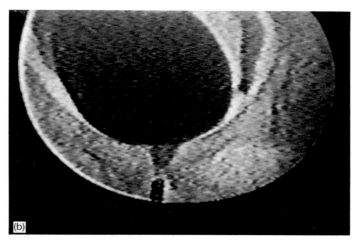

Fig 10.6. Detrusor/external sphincter dyssynergia. Radiograph at point A, high-pressure voiding with bilateral reflux (the dark rectangle at the bottom of the picture is the anal plug EMG prior to expulsion). C4–6 tetraplegia. (a) Cystometrogram; (b) still from cineoradiograph.

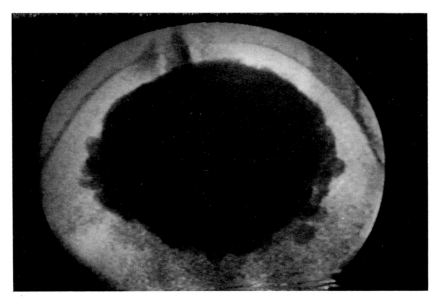

Fig. 10.7. High-pressure dyssynergia showing vesical diverticula and left-sided intraprostatic reflux.

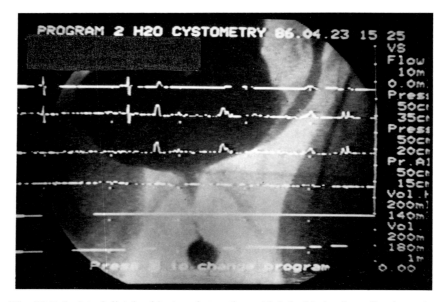

Fig. 10.8. Isolated distal sphincter obstruction with left-sided reflux. *Cauda equina* lesion.

Voiding by strain or compressive manœuvres may then have to be taught. The non-functioning urethra in the male does not often cause incontinence but frequently impedes micturition. Alpha-adrenergic blockade may prove useful in those where smooth muscle 'spasm' dominates, but more often transurethral sphincterotomy will be indicated despite the increased inconvenience this may entail. Although outlet obstruction due to the isolated distal sphincter is common in men (Fig. 10.8), obstruction may also occur in the female and is often mechanical due to bladder base descent on straining with kinking of the urethra. Due to the straining, there is often a significant degree of stress incontinence. In many patients a Stamey procedure or colposuspension is the treatment of choice, even though SIC may be required post-operatively.

Mixed lesions

A variety of patterns of neuropathy may occur as a result of lesions at the dorsolumbar junction. The principles outlined above need careful application to individual cases.

Artificial sphincters

The advent of a reliable artificial sphincter (Brantley Scott) has heralded an era of urological independence for many lower urinary tract invalids, but it is not suitable for all patients. Many surgeons have regarded being wheel-chair bound as an absolute contraindication to surgery, but, as always, there are exceptions. A urodynamic assessment is an essential prerequisite for consideration of an implant. The essentials have been described by Webster and Stephenson (1984) and others. Sphincter incompetence must be present or achieved by sphincterotomy. The bladder capacity must be adequate and without hypocompliance. Detrusor hyperreflexia must be of low pressure or capable of pharmacological control. Both detrusor contraction and compliance may be improved (i.e. reduced) by bladder replacement procedures (cystoplasty). The bladder must empty completely. Renal function should be good, with no or minimal vesicoureteric reflux, and the urine must be sterile. The operative field should be healthy and preferably unscarred. Finally, the patient should be sufficiently mobile, intelligent, and motivated to be able to use the device. It will be appreciated that more than one urodynamic study may be necessary to check that all these criteria have been achieved and will be required again if any problems persist after the implant. The possibility that SIC may still be required post-operatively must be borne in mind.

Conclusion

The videourodynamic studies outlined above are now accepted as standard practice in the assessment and treatment of the paraplegic and tetraplegic patient. It must not be forgotten that even in patients who show good recovery and resume walking, their lower urinary tract problem may and frequently does persist and evolve. Continued surveillance by these techniques is essential for an objective basis from which to make rational decisions about management with our present limited pharmacological and surgical armamentarium, and in order to establish the place of newer techniques as they are developed.

References

Abrams, P. H. (1977). The investigation of bladder outflow obstruction in the male. MD thesis, University of Bristol, Bristol, UK.

Awad, S. A. and Downie, J. W. (1977). Sympathetic dyssynergia in the region of the external sphincter: a possible source of lower urinary tract obstruction. *J. Urol.* **118**, 636–40.

Bary, P. R., Day, G., Lewis, P., Chawla, J., Evans, C., and Stephenson, T. P. (1982). Dynamic urethral function in the assessment of spinal injury patients. *Br. J. Urol.* **54**, 39–44.

Blaivas, J. G. (1982). The neurophysiology of micturition: a clinical study of 550 patients. *J. Urol.* **127**, 958–63.

Blaivas, J. G., Sinha, H. P., Sayed, A. A. H., and Labib, K. B. (1981). Detrusor–external sphincter dyssynergia. *J. Urol.* **125**, 542–4.

Bradley, W. E. (1972). Urethral electromyelography. *J. Urol.* **108**, 563–4.

Brown, M. and Wickham, J. E. A. (1969). The urethral pressure profile. *Br. J. Urol.* **41**, 211–17.

Denny-Brown, D. and Robertson, E. G. (1933). On the physiology of micturition. *Brain* **56**, 143–90.

Donker, P. J., Ivanovici, F., and Noach, E. L. (1972). Analysis of the urethral pressure profile by means of electromyography and the administration of drugs. *Br. J. Urol.* **44**, 180–93.

Eisen, A. and Elleker, E. (1980). Sensory nerve stimulation and evoked cerebral potentials. *Neurology* **30**, 1097–1105.

Gosling, J. A., Dixon, J. S. and Humpherson, J. R. (1983). Gross and microscopic anatomy of the urinary bladder. In *Functional anatomy of the urinary tract*, pp. 3.1–3.32. Gower Medical Publishing, London.

Hackler, R. H. (1977). A 25 year prospective mortality study in the spinal cord injured patient: comparison with the longterm living paraplegic. *J. Urol.* **117**, 486–8.

Iggo, A. (1955). Tension receptors in the stomach and the urinary bladder. *J. Physiol. (Lond.)* **128**, 593–607.

James, E. D. (1979). Continuous monitoring. *Urol. Clin. N. Amer.* **6**, 125–35.

Koizumi, K. and Brooks, C. M. (1974). The autonomic nervous system and its role in controlling visceral activities. In *Medical physiology* (13th edition) (ed. V. B. Mountcastle), pp. 783–812. C. V. Mosby Company, St Louis, Missouri.

Koyanagi, T. (1978). Denervation supersensitivity of the urethra to alpha adrenergics in the chronic neurogenic bladder. *Urol. Res.* **6**, 89–93.

Koyanagi, T., Arikado, K., Takamatsu, T., and Tsuji, I. (1982). Relevance of sympathetic dyssynergia in the region of external urethral sphincter: possible mechanism of voiding dysfunction in the absence of (somatic) sphincter dyssynergia. *J. Urol.* **127**, 277–82.

Krane, R. J. and Siroky, M. B. (1980). Studies on sacral-evoked potentials. *J. Urol.* **123**, 872–6.

Lapides, J., Friend, C. R., Ajemian, E. P., and Reus, W. F. (1962). Denervation supersensitivity as a test for neurogenic bladder. *Surg. Gynecol. Obstet.* **114**, 241–4.

Massey, J. A. and Abrams, P. (1985). Urodynamics of the lower urinary tract. *Clin. Obstet. Gynaecol.* **12**, 319–41.

McGuire, E. J. and Morrissey, S. G. (1982). The development of reflex bladder activity following spinal cord injury in cats and a method to control it. *Neurourol. Urodyn.* **1**, 211–20.

Nathan, P. W. (1956). Sensations associated with micturition. *Br. J. Urol.* **28**, 126–31.

Olsson, C., Siroky, M., and Krane, R. (1977). The phentolamine test in neurogenic bladder dysfunction. *J. Urol.* **117**, 481–5.

Rossier, A. B. and Ott, R. (1976). Bladder and urethral recordings in acute and chronic spinal cord injury patients. *Urol. Internat.* **31**, 49–59.

Sands, J. P., Constantineu, C. E., and Govan, D. E. (1972). Bladder pressure and its effect on mean arterial blood pressure. *Invest. Urol.* **10**, 14–18.

Schafer, W., Gerlach, R., and Hannapel, J. (1982). Microtip versus standard transducers: what do they really measure? *Proceedings of the XIIth annual meeting of the International Continence Society*, p. 53. ICS, London.

Sunder, G. S., Parsons, K. R., and Gibbon, N. O. K. (1978). Outflow obstruction in neuropathic bladder dysfunction: the neuropathic urethra. *Br. J. Urol.* **50**, 190–9.

Sundin, T., Dahlstrom, A., Norlen, L., and Svedmyr, N. (1977). The sympathetic innervation and adrenoceptor function of the human lower urinary tract in the normal state and after parasympathetic denervation. *Invest. Urol.* **14**, 322–8.

Taira, N. (1972). The autonomic pharmacology of the bladder. *Ann. Rev. Pharmacol. Toxicol.* **12**, 197–208.

Thomas, D. G. (1979). Clinical urodynamics in neurogenic bladder dysfunction. *Urol. Clin. N. Amer.* **6**, 237–53.

Vereecken, R. L. and Verduyn, H. (1970). The electrical activity of paraurethral and perineal muscles in normal and pathological conditions. *Br. J. Urol.* **42**, 457–63.

Webster, G. and Stephenson, T. P. (1984). Artificial urinary sphincters. In *Urodynamics, principles, practice and application* (ed. A. R. Mundy, T. P. Stephenson, and A. J. Wein), pp. 373–80. Churchill Livingstone, Edinburgh.

11

Pelvic floor reflexes and sphincter control systems in humans

Michael Swash

The S2, S3, and S4 segments of the spinal cord are especially concerned with pelvic floor function. The pelvic floor is a diaphragm of striated muscle forming the floor of the pelvis. It is perforated by the urethral and anal orifices and, in the female, by the vaginal orifice. The largest component of the pelvic floor consists of the paired *levator ani* muscles. These two muscles divide anteriorly to allow the urethral, vaginal, and anal outlets to penetrate the pelvic floor diaphragm. The lowermost fibres of the *levator ani* differ both by macroscopic and by microscopic criteria (Parks, Swash, and Urich 1977; Beersiek, Parks, and Swash 1979) from the main part of this muscle. The puborectalis muscle is a specialized sling of muscle arising from the pubis and passing backwards, posterior to the anorectum at the anorectal angle. Its uppermost fibres appear continuous posteriorly with those of the *levator ani* but, because of its pubic attachment, this muscle functions as a forward-pulling sling around the natural walls of the vagina, and the lateral and posterior walls of the anorectum. The histochemical fibre-type distribution within this muscle, and the fibre-size characteristics of the type-1 and type-2 fibres in this muscle, differ slightly from those of the *levator ani* and tend to resemble those of the external anal sphincter (Beersiek *et al.* 1979). The external anal sphincter muscle itself consists of a ring of striated muscle, the superficial component of which is attached to the skin of the anal margins, which encircles the anal canal just oral to the anal verge. This ring of muscle is innervated by the pudendal nerves (Kiff and Swash 1984*a*; Snooks and Swash 1986) whereas the puborectalis sling is innervated by direct motor branches of the S2 sacral root entering the muscle on its pelvic surface (Percy, Neill, Parks, and Swash 1981; Snooks and Swash 1986). The puborectalis muscle is sometimes termed the 'puboanal sling'. In the bladder the major component of the voluntary striated sphincter musculature, the periurethral striated sphincter muscle, encircles the urethra just below its site of exit from the bladder at the level of the urethra–vesical angle. An intramural component of the striated sphincter musculature of the urethra, like the puborectalis muscle, is innervated by direct pelvic branches from the sacral roots, whereas the periurethral striated sphincter is innervated by perineal branches of the pudendal nerves, and thus resembles the external anal sphincter in its embryological derivation (Snooks and Swash 1986). The histochemical characteristics of the periurethral and intramural components of the striated urethral sphincter muscles have been studied by Gosling (1979), who found that, as in the case of the external anal sphincter and puborectalis muscles, there was a predominance of type-1 fibres in these muscles.

The motoneurons that give rise to the innervation of the urethral and anal sphincter muscles also innervate the ischiocavernosus and bulbocavernosus muscles. They are derived from Onuf's nucleus (Onuf 1901) in the sacral spinal cord. There is controversy as to whether these neurons are all somatic efferent (Konishi, Sata, Mizuno, Itoh, Namura, and Sugimoto 1978; Mannen, Iwata, Toyokura, and Nagashima 1982), or whether they contain some visceral efferent neurons, thus giving rise to autonomic nerve fibres (Jacobsohn-Lask 1931; Rexed 1954; Sung 1982). Crowe, Light, Chilton and Burnstock (1985) have shown that the striated external urinary sphincter muscle receives vasopressin intestinal peptidergic innervation in addition to somatic efferent innervation, thus reopening this unresolved controversy.

In motor neuron disease, an acquired disease of the motor system in which there is rapid degeneration of anterior horn cells (Greenfield 1958), the urinary and anal sphincters are characteristically spared, both to clinical and histopathological examination (Mannen *et al.* 1982; Schroder and Reske-Nielsen 1984). However, the fate of the puborectalis muscle in this disease is not known. Since incontinence is not a feature of the disease, it would be expected that this muscle would be spared, but if it is phylogenetically part of the pelvicaudal *levator ani* muscle complex, although functionally active as a sphincter at the anorectal junction by virtue of the anal flap-valve mechanism (Parks 1975; Parks *et al.* 1977), it might be expected to be vulnerable to denervation in motor neuron disease.

The *levator ani* muscle, but not the puborectalis or external anal sphincter muscles, shows unusual sexual dimorphism in the size relationships of its muscle fibres. In women, the type-1 fibres in this muscle are markedly larger than the type-2 fibres (Beersiek *et al.* 1979), an observation at variance with that in all other human striated muscles (Polgar, Johnson, Weightman, and Appleton 1973). This may represent a hormone-dependent phenomenon, and this feature may therefore be important in understanding the frequency and pathogenesis of pelvic floor disorders in women. In the rat, part of this muscle undergoes involution during the menarche (Cihak, Guttman, and Hanzlikova 1970). There are thus a number of ill-understood features of the pelvic floor muscles that merit further anatomical study.

Continence and the pelvic floor sphincters

These voluntary muscular systems are important in the maintenance of continence (Henry and Swash 1985). The functional interrelations between the striated and the non-striated muscular systems in the pelvic sphincter regions is currently not well understood, but it is clear that there must be close functional patterns of activity and relaxation between the smooth and striated muscular systems in order to achieve normal continence and defaecation or micturition. Indeed, there are close parallels in the results of physiological investigations of the bladder–urethral and rectal–anal relationships. Anorectal manometry has shown that there is a zone of pressure in the anal canal in the resting state of about 60–100 cm of water. This is present in the anal sphincter region and results largely from resting activity in the internal anal sphincter muscle (Bennett and Duthie 1964; Wheatley, Hardy, and Dent 1977). Bilateral pudendal nerve block results in a fall of this resting pressure of only about 15 per cent, this value representing the contribution of the external anal sphincter to resting pressure in the sphincter zone (Frenckner and Ihre 1976). The precise role of this internal anal sphincter-generated pressure zone in the anal canal in the maintenance of normal faecal incontinence is uncertain, but it is possible that it is important not so much in maintaining continence as in modulating sensory input in relation to anal canal contents. For example, division of the smooth, involuntary internal sphincter muscle during the treatment of anal fissure does not result in incontinence, although there is often some difficulty in relation to retention of liquid stool or of large quantities of *flatus*. In addition, division of the external anal sphincter similarly does not result in incontinence (Milligan and Morgan 1934).

The major muscle of faecal continence is believed to be the puborectalis muscle sling. Electromyographic (EMG) studies of both the external anal sphincter and puborectalis muscles have shown that these two muscles are in a state of continuous basal activity, even during sleep (Floyd and Walls 1953; Parks, Porter, and Melzack 1962). These muscles contain muscle spindles, and Bishop (1959) showed that dorsal root section, in the cat, abolished this tonic activity in these muscles, illustrating their dependence on a spinal reflex arc. A major aspect of idiopathic (neurogenic) faecal incontinence is widening of the normal anorectal angle, the angle made between the anal canal and the rectum by the tonic pull of the puborectalis muscle. When this angle is flattened abdominal compression, straining, or coughing may result in the passage of faeces from the rectum into the anal canal and thus to involuntary loss of faecal material, constituting incontinence (Parks, Porter, and Hardcastle 1966; Parks 1975; Bartram and Mahieu 1985). This mechanism for the maintenance of faecal continence was termed a flap-valve effect by Parks (1975). It has been pointed out that this is a peculiarly effective sphincter mechanism since the greater the intra-abdominal force the more secure the flap-valve mechanism (see Henry and Swash 1985).

Reflex control mechanisms of pelvic floor sphincters
In the maintenance of continence various reflex mechanisms play a key role. The normal state of continence is unconscious, yet effective. It is interrupted by defaecation or micturition but, to some extent, the timing of these evacuation mechanisms can be controlled. Although the puborectalis and external anal sphincter muscles are in a state of tonic contraction, mediated reflexly as discussed above, this is modulated by activity. For example, a sudden cough causes a transient increase in the intra-anal pressure in the sphincter zone by as much as 200 cm of water. This reflects a brisk increase in external anal sphincter activity induced reflexly by the cough. The puborectalis muscle similarly contracts in response to a cough and these responses are termed 'the cough reflex'. During other activities, such as talking and changes in posture, reflex contraction also develops in the puborectalis and external anal sphincter muscles, as a transient response, so maintaining the anorectal angle and preventing incontinence (Taverner and Smiddy 1959; Parks *et al.* 1962). On the other hand, sudden distension of the rectum, for example, by a faecal bolus or by an experimental balloon inserted through the anal canal, results in marked inhibition of contraction of the smooth internal anal sphincter muscle and of the striated external anal sphincter muscle. These two components of the response are jointly termed the rectoanal relaxation reflex. This response was described by Gowers (1877) and later studied by Denny-Brown and Robertson (1935). The major component consists of internal anal sphincter relaxation. This latter response to rectal distension disappears after circumferential rectal myotomy between the balloon and the internal anal sphincter muscle, and is also absent after rectal excision although, following coloanal anastomosis the response reappears in some patients after a period of several months. Pharmacological blockade of the sympathetic innervation to the internal anal sphincter muscle, by procainization of the pre-sacral nerves intra-operatively, did not abolish the rectoanal relaxation reflex (Lubowski, Nicholls, Swash, and Jordan 1987), an observation further suggesting that this response is the result of a locally mediated reflex through rectoanal ganglia in the wall of the viscus. This rectoanal relaxation reflex is important in the passage of a faecal bolus along the rectum and anal canal during defaecation since it allows the easier passage of the bolus by dilatation of the anal orifice in front of the slowly moving bolus. Clearly, the voluntary external anal sphincter must be part of this response, as was suggested by Gowers (1877). Observation of the act of defaecation during coloproctography in humans, and of defaecation in other species, shows that the anal ring dilates to allow the passage of the bolus during the act of defaecation.

The *closing reflex* consists of a rapid burst of electrical activity in the external anal sphincter and pelvic floor muscles immediately following defaecation. This is accompanied by reconstitution of the anorectal angle, and of the position of the pelvic floor and is followed by a reactivation of resting tonic activity in the puborectalis and external anal sphincter muscles. Clinical and radiological observations suggest that the internal anal sphincter similarly undergoes a return of tonic contraction following defaecation.

The patterns of activity in the external anal sphincter and puborectalis muscles can be studied by electromyographic monitoring. By these means the cough reflex, the activity of these muscles during defaecation proctography (Womack, Williams, Holmfield, Morrison, and Simpkins 1987), and the patterns of activity generated during straining, as if during defaecation and during sphincter squeeze exercises, can be assessed. Parks *et al.* (1962) and Porter (1962) showed that contraction of the puborectalis and external anal sphincter muscles is inhibited in patients requested to strain or bear down as if preparing for defaecation. It is possible that this inhibition of contraction is reflexly mediated by a spinal cord response to the pressure generated in the rectum during bearing-down. During the early phase of defaecation the distal rectum opens out like a funnel. This relaxation of the rectum extends into the upper part of the anal canal as it fills with faeces during the process of evacuation. The critical initiating factors in this chain of events are not understood.

During defaecation propagated waves, in some respects resembling peristalsis, have been seen in the rectum, suggesting that smooth, non-striated muscle in the walls of the rectum is involved in the reflex activity making up the defaecatory act.

Defaecation can be induced by a number of mechanisms. First, it commonly occurs as part of the gastrocolic reflex, following the ingestion of a meal, particularly in the morning. Second, it is a truism that most persons can voluntarily initiate defaecation, although it is by no means certain that the sequence of activation of the striated and non-striated muscular systems during voluntarily initiated defaecation is the same as that during reflexly initiated defaecation, as in the gastrocolic reflex. Third, temporal factors are important. In most people defaecation occurs at a particular time of day, as is shown by the effect of jet travel on such time-locked reflexes.

Disorders of faecal continence

With lesions in the nervous system a number of abnormalities in the patterns of defaecation, sometimes amounting to incontinence, may develop. For example, in patients with widespread damage to the brain, as in anoxic encephalopathy, progressive dementias, or other degenerative diseases, or multi-infarct disease, there is loss of social control. This has commonly been associated with frontal lobe lesions. The patient with frontal lobe damage may defaecate or urinate in inappropriate situations. However, the actual processes of defaecation and of micturition appear normal in such subjects. They are not so much incontinent as have lost social control so that they are prepared to defaecate or micturate in socially unacceptable situations. In the natural history of a progressive dementia, as the disorder becomes more severe the pattern of abnormality of sphincter control becomes more complex and defaecation and micturition become uncontrolled. Incontinence of urine and faeces results from bilateral brain-stem lesions, and from lesions in the spinal cord, when the descending pathways controlling the sphincter musculature, which travel in the medial aspects of the corticospinal system, are damaged. Thus intrinsic lesions within the brain stem and spinal cord, such as multiple sclerosis, infarction, or tumour, often present with bladder and bowel disturbances, whereas extrinsic compression of the brain stem or spinal cord results in retention or incontinence of urine and faeces only relatively late in the course of the disorder.

Investigation of pelvic floor sphincter muscles

The corner-stone of clinical investigation of the pelvic floor sphincters remains the clinical history and the physical examination. These techniques are powerful and capable of localizing the functional abnormality. Information obtained by anorectal manometry, or by its related investigation in the bladder, the cystometrogram, supplements the results of clinical assessment. In the case of anorectal manometry the examiner's finger in the rectum can provide strictly comparable information, although without absolute quantification. Such clinical techniques are not available for the urethra and the cystometrogram is therefore an important part of the urological examination. The cystometrogram has been refined to include videocystometrogram assessments with manometric assessment, and urethral pressure profile measurements in order to examine the pressure zone at the upper level of the urethral in the sphincter zone.

The function of the striated sphincter musculature of the urethra and of the anorectum can also be assessed by electromyography and by examination of function in the efferent motor nerves innervating the sphincter muscles. These investigations have been particularly applied to the study of incontinence, but have also been found useful in the assessment and management of other pelvic floor disorders, such as constipation, solitary rectal ulcer syndrome, perineal descent syndromes, and anal pain, and in the pre-operative assessment of patients undergoing pelvic floor surgery. The methods we have devised for the investigation of pelvic floor disorders have been described in detail elsewhere (Snooks, Barnes, and Swash 1984a;

Snooks, Swash, Setchell, and Henry 1984*b*; Henry and Swash 1985; Snooks and Swash 1985*a*, 1985*b*).

Transcutaneous spinal stimulation
Direct electrical stimulation of the *cauda equina* is achieved using a modification (Snooks and Swash 1985*a–c*) of the technique devised by Merton, Hill, Morton, and Marsden (1982). The patient is placed in the left lateral position and a separate ground electrode is connected from the right upper thigh to the pre-amplifier of the Medelec MS6 EMG apparatus. A single impulse of 500 to 1500 V, of 0.5 ms duration, and decaying with a time constant of 50 μs, is delivered through two saline-soaked pad electrodes with the cathode applied to the skin at the level of the spine of the first lumbar vertebra. The anode is directed cranially. An initial 200 V stimulus is applied and increased in 200 V increments until the amplitude and latency of the evoked muscle response in the external anal sphincter or puborectalis muscles do not change with further stimulus increments, thus indicating that the stimulus is supramaximal. Application of the same stimulus to the spine at the level of the fourth lumbar vertebral spine enables a latency between the L1 and L4 levels to be measured, thus allowing calculation of motor conduction and velocity in the motor roots of the *cauda equina* (Swash and Snooks 1986).

External anal sphincter response   The response in the external anal sphincter muscle is recorded through two poles of a 3 cm long, three-pole telephone jack-plug surface electrode lubricated with electrode jelly, placed in the anal canal and connected to ground. The onset of the stimulus triggers the oscilloscope of the EMG apparatus. The latency of the responses following five consecutive supramaximal stimuli delivered at intervals is measured from the onset of the stimulus to the onset of the response on the paper printout by two independent observers.

Puborectalis response   Using the same technique of transcutaneous spinal stimulation the latency of the response to the puborectalis muscle was measured. The compound muscle action potential evoked in the puborectalis muscle was recorded using a rubber finger stall with two 1 cm diameter circular, steel-surface electrode-plates, placed 1 cm apart, positioned at its tip (Kiff and Swash 1984*a,b*; Henry, Snooks, Barnes, and Swash 1985; Snooks, Swash, and Henry 1985). The finger bearing this device was inserted into the rectum so that the recording surfaces were in contact with the puborectalis muscle bar; i.e. the recording surfaces faced posteriorly. The contraction response of the puborectalis muscle, which could be palpated with the finger, was displayed on the EMG apparatus and the latency measured in the same way as that for the external anal sphincter muscle.

Urethral striated sphincter response The response of this muscle to the same spinal stimulus was recorded using a pair of intraurethral surface electrodes mounted on a sterilizable Foley catheter (Snooks and Swash 1985*c*).

Pudendal and perineal nerve terminal motor latencies
These methods were developed from the technique of electroejaculation described for use in men with impotence due to paraplegia by Brindley (1981). These methods have been described in detail elsewhere (Kiff and Swash 1984*a*; Snooks and Swash 1985*b*). The pudendal nerve can be stimulated near the ischial spine through the wall of the rectum using a pair of electrodes mounted on the tip of a rubber finger stall similar to that used for clinical examination of the rectum. Two steel-surface electrodes mounted at the base of the finger stall, 3.5 cm from the cathodal stimulating electrode, are used to record the response in the external sphincter muscle. The latter can be felt by the examiner and the tip of the finger stall moved slightly in the rectum to achieve the best response. Square-wave stimuli of 0.1 ms duration and up to 50 V are given at 1 s intervals. The same stimulus can be used to record the compound muscle action potential evoked in the periurethral striated sphincter muscle, using the catheter-mounted electrode described above. Latencies to supramaximal responses are measured on paper printouts of these responses (Snooks and Swash 1985*b,c*) and, since the apparatus is standardized, a set of normal values for the pudendal (PNTML) and perineal (PerNTML) nerve terminal motor latencies has been obtained (Kiff and Swash 1984*a,b*; Snooks *et al.* 1984*a*; Snooks and Swash 1985*b,c*).

These electrophysiological studies have allowed separate evaluation of the anal and urinary sphincter musculature and their innervation. During transrectal stimulation of the pudendal nerve, to measure the PNTML by recording the evoked compound muscle action potential in the external anal sphincter in control subjects, no contraction of the puborectalis is felt by the examiner's finger. This suggests that the puborectalis muscle is not activated by electrical stimulation of the pudendal nerves at the level of the ischial spines. The latencies of the compound muscle action responses in the external anal sphincter and periurethral striated sphincter muscles, which are both innervated by the pudendal nerves, differ by 0.5 ± 0.2 ms, a difference consistent with the additional length of the perineal branch of the pudendal nerve innervating the latter muscle (approximately 2.5 cm at a nerve conduction velocity of 50 m/s).

Spinal stimulation at the level of the L1 vertebral spine produced recordable muscle action potentials in the puborectalis, external anal sphincter, and urethral striated sphincter muscles. The difference in latency between the compound muscle action potentials evoked in the external anal sphincter and in the urethral striated sphincter muscles after spinal stimulation at L1 was 0.6 ± 0.2 ms. The compound muscle action potential recorded in the puborectalis muscle after this stimulus was evoked 0.7 ± 0.4 ms *earlier* than that in the external anal sphincter muscle, suggesting that the innervation of the puborectalis from the point of stimulation at L1 was about 3.5 cm shorter than that to the external anal sphincter (assuming a conduction velocity of 50 m/s) (Snooks and Swash 1986).

Observations on incontinent patients

In patients with faecal incontinence the PNTML is increased, a finding consisting with a lesion in the distal part of the innervation of the external anal sphincter muscle (Kiff and Swash 1984a,b). In patients with double incontinence the PNTML was increased to a similar extent, but in these patients the PerNTML was also increased, a finding suggestive of additional damage to this innervation (Snooks and Swash 1985b,c). Spinal stimulation at the L1 level in these patients with double incontinence showed that, although the latency of the evoked compound muscle action potentials in the striated urethral sphincter musculature was also usually increased, in seven (41 per cent) patients it was normal, although the PerNTML in these patients was greatly increased. This observation suggests that there must be a dual innervation to this sphincter complex, and that in these patients only the perineal component was abnormal. In patients with genuine stress urinary incontinence, the PerNTML was increased, but the PNTML was normal (Snooks *et al.* 1985).

Latencies to the external anal sphincter and puborectalis muscles from spinal stimulation showed increases of a similar degree. The increased latency to the external anal sphincter muscle from L1 spinal stimulation is largely due to slowed conduction in the distal part of the pudendal innervation of this muscle, as shown by the increase in the PNTML (see discussion in Kiff and Swash 1984a,b; Snooks *et al.* 1984a; Snooks and Swash, 1985b).

Application to spinal cord injury

Trauma to the spinal cord presents two major problems, that of damage to the spinal nerve roots, particularly of the *cauda equina*, and of damage to the spinal cord itself. In some patients there is probably injury both to the *cauda equina* nerve roots and to the lower part of the spinal cord. Diagnosis of these two aspects of spinal cord injury is important because of the implication for management in the acute phase, and the possibility that a *cauda equina* lesion might respond to treatment, particularly surgical decompression, if recognized in the early stages. The electrophysiological methods outlined above may be used for diagnosis of these problems.

Cauda equina lesions

Adequate methods for electrophysiological diagnosis of *cauda equina* lesions can improve management by providing evidence to support the clinical diagnosis and thus to justify interventional investigation at an appropriate time. Conventional methods, consisting of electromyography in radicular distributions, thus including spinal extensor muscles, an assessment of the H-reflex latency in the L5/S1 muscles, and of F-wave latencies in a wider distribution of nerve roots, have been used with some success. In addition, somatosensory-evoked potential studies are often used for this purpose. However, the major practical problem in the assessment of patients with *cauda equina* trauma, as in other *cauda equina* lesions such as lumbosacral canal stenosis or lumbosacral disc disease is the extent of motor involvement. For this purpose we have developed the technique of transcutaneous stimulation of the spinal roots, which we introduce for assessment of proximal motor conduction in patients presenting with incontinence. The method is described briefly above. It consists of measuring the motor latencies to the external anal sphincter or puborectalis muscles from L1 and L4 vertebral levels using transcutaneous spinal stimulation (500–1500 V, time constant 50 μs, pulse duration 50 μs). The method can be adapted to assessment of other root distributions by placing suitable surface electrodes on muscles representing other lumbosacral roots, for example quadriceps, *gastrocnemius/soleus* muscles, and extensor *digitorum brevis* muscles. The motor latencies to these roots from the spinal stimulation at the two levels can be assessed in comparison on the two sides, or by reference to a control population, corrected for height. We have ourselves preferred to use latencies to the pelvic floor sphincter muscles in problems of this type because these latencies allow assessment of conduction through the full length of the spinal cord and through the S2–4 motor roots, whereas latency measurements to other muscles in the legs omit the sacral spinal cord and sacral roots from the estimation. In these measurements we have preferred to use the ratio of the L1:L4 motor latencies rather than the absolute latencies since this ratio provides an in-built correction for height. The L1:L4 latency ratio to the puborectalis was 1.36 ± 0.09 (SD) in control subjects and 1.72 + 0.13 (SD) in patients with *cauda equina* lesion (Swash and Snooks 1986). The pudendal nerve terminal motor latency was normal in most of these patients. The single-fibre EMG fibre density in the muscles of the pelvic floor was increased in the patients with *cauda equina* lesions but, since this may be increased also in patients with pelvic floor neuropathy associated with lesions within the pelvis, or in the pelvic floor for other causes, this is not a reliable measurement. These electrophysiological methods are attractive since they may be used not only for the selection of patients for myelography, but also to follow progress in patients managed conservatively.

Motor conduction velocity in the human spinal cord

We have used transcutaneous electrical stimulation in the central nervous system to measure motor conduction velocity in the human spinal cord in control subjects and in patients with neurological disease. The motor conduction velocity between the C6 and L1 vertebral levels was 67.4 ± 9.1 ms, probably representing maximal conduction velocity in the corticospinal tracts. Motor conduction velocity in the *cauda equina* between L1 and L4 vertebral levels in the same subjects (see above) was 57.9 ± 10.3 ms (Snooks and Swash 1985a). We have found that motor conduction velocity in the spinal cord may be slowed in patients with multiple sclerosis, when this is associated with corticospinal abnormalities in the legs, and similar slowing was also found in a patient with radiation myelopathy. In addition, in a subsequent series of investigation we found some slowing of motor conduction velocity in the spinal cord in patients with motor neuron disease, although to a lesser degree than in multiple sclerosis or other cord lesions associated with demyelination (Ingram and Swash 1986; Ingram, Thompson, and Swash 1987). We have no experience, as yet, of this technique in patients with spinal cord injury but would expect that it should provide a useful method in assessing the integrity of the fast-descending motor pathways in the spinal cord, particularly in patients with partial cord lesions. The end-point for recording may consist, as in our studies, of pelvic floor sphincter muscles, in which case the whole sacral cord is involved in the assessment, or surface electrodes may be placed over limb muscles, representing lumbosacral root innervations, in which case the sacral cord is excluded from the measurement. The method may be combined with measurement of central motor conduction time from cortex to C6, thus allowing motor conduction to be assessed accurately in the whole length of the motor system from the cerebral cortex to the peripheral nerve. In addition to this simple application the method is of considerable interest in relation to our understanding of motor neuron excitability, a problem of relevance to the assessment of partial and complete cord lesions.

Acknowledgements

I would like to take this opportunity to thank the St Mark's Hospital Research Foundation, the Royal College of Surgeons of England, the Sir Alan Parks Research Fund of the Royal College of Surgeons of England, the

London Hospital Special Trustees, and the various local research funds supporting the research fellows who have come to work on the disorders of the pelvic floor at St Mark's during the last nine years, for their generous support of this work.

References

Bartram, C. and Mahieu, P. H. G. (1985). Radiology of the pelvic floor. In *Coloproctology and the pelvic floor* (ed. M. M. Henry and M. Swash), pp. 151–86. Butterworth, London.

Beersiek, F., Parks, A. G., and Swash, M. (1979). Pathogenesis of anorectal incontinence; a histometric study of the anal sphincter musculature. *J. Neurol. Sci.* **42**, 111–27.

Bennett, S. C. and Duthie, H. L. (1964). The functional importance of the internal anal sphincter. *Br. J. Surg.* **51**, 355–7.

Bishop, B. (1959). Reflex activity in the external anal sphincter of the cat. *J. Neurophysiol.* **22**, 679–92.

Brindley, G. S. (1981). Electroejaculation; its technique, neurological implications and uses. *J. Neurol. Neurosurg. Psychiat.* **44**, 9–18.

Cihak, R., Guttman, E., and Hanzlikova, V. (1970). Involution and hormone-induced persistence of the M sphincter (levator) ani in female rats. *J. Anat. (Lond.)* **106**, 93–101.

Crowe, R., Light, J. K., Chilton, C. P., and Burnstock, G. (1985). Vasoactive intestinal polypeptide (VIP)-immunoreactive nerve fibres associated with the striated muscle of the external urethral sphincter. *Lancet* **i**, 47–8.

Denny-Brown, D. and Robertson, G. (1935). An investigation of the nervous control of defaecation. *Brain* **58**, 256–310.

Floyd, W. F. and Walls, E. W. (1953). Electromyography of the sphincter ani externus in man. *J. Physiol.* **122**, 599–609.

Frenckner, B. and Ihre, T. (1976). The influence of autonomic nerves on the intrenal anal sphincter in man. *Gut* **17**, 306–12.

Gosling, J. (1979). The structure of the bladder and urethra in relation to function. *Urol. Clin. N. Amer.* **6**, 31–8.

Gowers, W. R. (1877). The automatic action of the sphincter ani. *Proc. R. Soc. (Lond.)* **26**, 77–84.

Greenfield, J. G. (1958). *Neuropathology*. Edward Arnold, London.

Henry, M. M. and Swash, M. (1985). *Coloproctology and the pelvic floor*. Butterworth, London.

Henry, M. M., Snooks, S. J., Barnes, P. R. H., and Swash, M. (1985). Investigation of disorders of the anorectum and colon. *Ann. R. Coll. Surg. (England)* **67**, 355–60.

Ingram, D. A. and Swash, M. (1987). Central motor conduction is abnormal in motor neuron disease. *J. Neurol. Neurosurg. Psychiat.* **50**, 159–66.

Ingram, D. A., Thompson, A., and Swash, M. Abnormalities in central motor conduction in multiple sclerosis as revealed by magnetic stimulation of the brain. *J. Neurol. Neurosurg. Psychiat.* (in press).

Jacobsohn-Lask, J. (1931). Uber den medialen Sympatikuskern des menschlichen Ruckenmarks. *Z. f. d. ges. Neurol. Psychiat.* **134**, 644–56.

Kiff, E. S. and Swash, M. (1984a). Slowed conduction in the pudendal nerves in idiopathic (neurogenic) faecal incontinence. *Br. J. Surg.* **71**, 614–16.

Kiff, E. S. and Swash, M. (1984b). Normal proximal and delayed distal conduction in the pudendal nerves of patients with idiopathic (neurogenic) faecal incontinence. *J. Neurol. Neurosurg. Psychiat.* **47**, 820–3.

Konishi, A., Sata, M., Mizuno, N., Itoh, K., Namura, S., and Sugimoto, T. (1978). An electron microscopic study of Onuf's nucleus in the cat. *Brain Res.* **156**, 333–8.

Lubowski, D. Z., Nicholls, R. J. Swash, M., and Jordan, M. J. (1987). Neural control of internal and sphincter function. *Br. J. Surg.* **74**, 668–70.

Mannen, T., Iwata, M., Toyokura, Y., and Nagashima, K. (1982). The Onuf's nucleus and the external anal sphincter muscles in ALS and Shy-Drager syndrome. *Acta Neuropathol. (Berlin)* **58**, 255–60.

Merton, P. A., Hill, D. K., Morton, H. B., and Marsden, C. D. (1982). Scope of a technique for electrical stimulation of the human brain, spinal cord and muscle. *Lancet* **ii**, 597–600.

Milligan, E. T. C. and Morgan, C. N. (1934). Surgical anatomy of the anal canal with special reference to anorectal fistulae. *Lancet* **ii**, 1150–6.

Onuf, B. (1901). On the arrangement and function of the cell groups of the sacral region of the spinal cord in man. *Arch. Neurol. Psychopathol. (Chicago)* **3**, 387–412.

Parks, A. G. (1975). Anorectal incontinence. *Proc. R. Soc. Med.* **68**, 681–90.

Parks, A. G., Porter, N. H., and Hardcastle, J. D. (1966). The syndrome of the descending perineum. *Proc. R. Soc. Med.* **59**, 477–82.

Parks, A. G., Porter, N. H., and Melzack, J. (1962). Experimental study of the reflex mechanism controlling the muscles of the pelvic floor. *Dis. Colon Rectum* **5**, 407–14.

Parks, A. G., Swash, M., and Urich, H. (1977). Sphincter denervation in anorectal incontinence and rectal prolapse. *Gut* **18**, 656–65.

Percy, J. P., Neill, M. E., Parks, A. G. and Swash, M. (1981). Electrophysiological study of motor nerve supply of pelvic floor. *Lancet* **i**, 16–17.

Polgar, J., Johnson, M. A., Weightman, D., and Appleton, D. (1973). Data on fibre size in thirty-six human muscles; an autopsy study. *J. Neurol. Sci.* **19**, 307–18.

Porter, N. H. (1962). A physiological study of the pelvic floor in rectal prolapse. *Ann. R. Coll. Surg. (England)* **31**, 379–404.

Rexed, B. (1954). A cytoarchitectonic atlas of the spinal cord in the cat. *J. Comp. Neurol.* **100**, 297–379.

Schroder, H. D. and Reske-Nielsen, E. C. (1984). Fiber types in the striated urethral and anal sphincters. *Acta Neuropathol. (Berlin)* **60**, 278–82.

Snooks, S. J. and Swash, M. (1985a). Motor conduction velocity in the human spinal cord: slowed conduction in multiple sclerosis and radiation myelopathy. *J. Neurol. Neurosurg. Psychiat.* **48**, 1135–9.

Snooks, S. J. and Swash, M. (1985b). Nerve stimulation techniques. In *Coloproctology and the pelvic floor* (ed. M. M. Henry and M. Swash), pp. 112–24. Butterworth, London.

Snooks, S. J. and Swash, M. (1985c). Perineal nerve and transcutaneous spinal stimulation: new methods for the investigation of the urethral striated sphincter musculature. *Br. J. Urol.* **56**, 406–9.

Snooks, S. J. and Swash, M. (1986). The innervation of the muscles of continence. *Ann. R. Coll. Surg. (England)* **68**, 45–9.

Snooks, S. J., Barnes, P. R. H., and Swash, M. (1984a). Damage to the innervation of the voluntary anal and periurethral striated sphincter musculature in incontinence; an electrophysiological study. *J. Neurol. Neurosurg. Psychiat.* **47**, 1269–73.

Snooks, S. J., Swash, M., and Henry, M. M. (1985). Abnormalities in central and peripheral nerve conduction in anorectal incontinence. *J. R. Soc. Med.* **78**, 294–300.

Snooks, S. J., Swash, M., Setchell, M., and Henry, M. M. (1984b). Injury to innervation of pelvic floor sphincter musculature in childbirth. *Lancet* **ii**, 546–50.

Sung, J. H. (1982). Autonomic neurons of the sacral spinal cord in ALS, anterior poliomyelitis, and intranuclear hyaline inclusion disease; distribution of sacral autonomic neurons. *Acta Neuropathol. (Berlin)* **56**, 233–7.

Swash, M. and Snooks, S. J. (1986). Slowed motor conduction in lumbosacral nerve roots in cauda equina lesions; a new diagnostic technique. *J. Neurol. Neurosurg. Psychiat.* **49**, 808–16.

Taverner, D. and Smiddy, F. G. (1959). An electromyographic study of the external anal sphincter and pelvic diaphragm. *Dis. Colon Rectum* **2**, 153–60.

Womack, N. R., Williams, N. S., Holmfield, J. H. M., Morrison, J. F. B., and Simpkins, K. C. (1987). New method for the dynamic assessment of anorectal function in constipation. *Br. J. Surg.* **85**, 994–8.

Wheatley, I. C., Hardy, K. J., and Dent, J. (1977). Anal pressure studies in spinal patients. *Gut* **18**, 488–90.

Yamamoto, T., Satomi, H., Ise, H., Takamata, H., and Takahashi, K. (1978). Sacral spinal innervation of the rectal and vesical smooth muscles and the sphincteric striated muscles as demonstrated by the horseradish peroxidase method. *Neurosci. Lett.* **7**, 41–7.

Clinical neurophysiological evaluation of motor and sensory functions in chronic spinal cord injury patients

Milan R. Dimitrijevic

Introduction

The spinal cord integrated with the rest of the central nervous system is highly involved in mediating and processing sensory input and the pre-final and final execution of motor control. Spinal cord injury causes different degrees of structural deterioration and impairment of functions. After the acute stage when dysfunction is more extensive, the condition stabilizes and the dysfunction can be described clinically as transected, incomplete lesion, diffuse, or hemisected. Another clinical syndrome is caused by the lesion when it involves completely or partially the posterior or the anterior structure of the spinal cord.

Thus the neurological assessment of spinal cord dysfunction is based on the analysis of disintegration between the ascending and descending motor systems. It is a relatively simple task to examine whether sensation for touch, pain, cold, heat, position, and two points of discrimination are preserved, impaired or completely absent. The same applies to the motor system to show if volitionally induced skilful or gross movements are present or impaired and, when impaired, how these functions affect the patient's posture and ability to stand and walk. However, in chronic spinal cord lesions it is not only a matter of conducting an inventory of altered or normal ascending and descending spinal cord functions, but also of being able to describe, to identify, and to prove the newly established and generated sensory–motor functional interaction between the injured portions of the spinal cord. It is reasonable to assume that in chronic spinal cord lesions a series of pathological changes have been established and maintained for months and years and as such they represent a 'new anatomy'. The questions then arise as to how to assess this new condition and how to determine if altered sensation and motor activity have organization and control mechanisms which can be used to minimize the consequences of the injury and to augment residual control.

In this chapter I will outline the clinical neurophysiological evaluation of motor and sensory functions of the chronically injured spinal cord and will illustrate these methods and their practical usefulness.

Motor activity

Studies of motor activity can be based on recordings of mechanical events of volitionally, automatically, or reflexly induced movements or simultaneous electromyographic recordings of motor unit activity of different muscle groups. The latter method has the advantage as well as the disadvantage of recording electrical activity regardless of whether the movement elicited is isometric or isotonic. We have found it practical and informative to use polyelectromyographic recordings, recording from ten to sixteen muscle groups simultaneously while applying different manœuvres to elicit reflex, automatic, and volitional motor activity (Dimitrijevic and Sherwood 1980). The protocol we are using consists of recordings of segmental reflex activity, tendon jerks, clonus, vibratory tonic reflex, manually induced passive tonic stretch reflex, plantar reflex, and cutaneomuscular reflexes. It tests the patient's ability to fully relax during the activation of motor units for ten minutes, the effects of reinforcement manœuvre on motor unit activation, and volitional activation of gross and fine movements. Recordings are made with surface electrodes and an amplifier and are recorded on a jet ink recorder (Dimitrijevic 1984).

This systematic polyelectromyographic recording procedure has led us to conclude that such recordings of motor activity in these two distinct clinical groups of complete and incomplete spinal cord injury patients can provide subclinical evidence of brain influence, even below the injured portion of the spinal cord, and clinical findings of complete lesion (Dimitrijevic and Faganel 1985). Polyelectromyographic evaluation of motor activity has also been found useful in patients with clinically incomplete lesions to differentiate between the organized activity of motor units and the co-activation of disorganized motor activity and its disorganization between segmental and suprasegmental activity (Malezic, Dimitrijevic, Dimitrijevic, and Sherwood 1984).

On the basis of these studies of motor control, we have been able to describe the characteristic features of neurocontrol in clinically complete and incomplete patients. These findings are summarized in Fig. 12.1, where it is shown that in clinically complete paraplegia, with no clinical or subclinical signs of brain influence below the level of the lesion, segmental activity has a phasic and unsustained character. In the clinically complete lesion, the presence of tonic and sustained features of segmental activity or stereotyped flexor- or extensor-induced movement are signs of subclinical evidence of brain influence below the level of the lesion. Further improve-

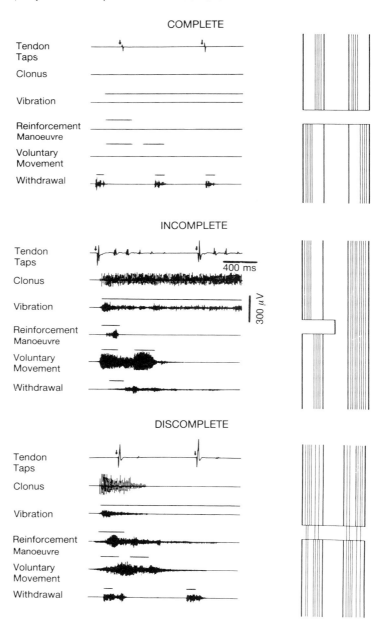

Fig. 12.1. On the left there are three typical polyelectromyographic findings of segmental and suprasegmental activity in patients with complete, discomplete (clinically complete but physiologically incomplete), and incomplete chronic chronic spinal cord lesion.

ment of suprasegmental neurocontrol will lead to the clinical recognition of the incomplete spinal cord lesion, the appearance of volitionally induced flexor and extensor gross movement, and to postural control, and fine skilful volitional activity and gait.

Similar polyelectromyographic recordings from several leg and trunk muscles or even upper limbs can also be used for the assessment of upper motor neuron functions in ambulatory spinal cord injury patients.

By means of polyelectromyographic recordings of leg muscles during gait with foot switches, indicating stance and swing phase, it is possible to measure the duration of different gait phases and compare them to the pattern of neurocontrol. These studies are useful in documenting changes in gait due to spontaneous recovery or induced recovery. The characteristic gait features of several ambulatory spinal cord injury patients are shown in Fig. 12.2.

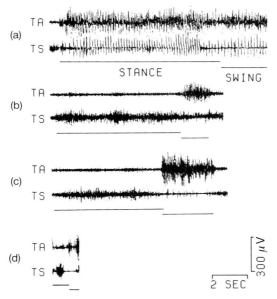

Fig. 12.2. Illustration of the impaired equilibrium between the ankle flexors and extensors during the stance and swing phase in ambulatory spinal cord injury patients. The most severe impairment of gait is shown in (a) and the least in (c). In comparison, (d) is the scale of a swing and stance phase of a neurologically healthy subject.

Recording of muscle tone

The majority of chronic spinal cord injury patients also suffer from increased muscle tone during passive stretch, which can be short- or long-lasting, phasic, or tonic. The simplest method to measure muscle hypertonia is to mount a goniometer at the ankle, which is passively flexed and extended, and to measure the rate of stretch through the application of constant force. This technique, known as the pendulum test, can also be upgraded by simultaneous recording of electromyographic (EMG) activity from the biceps and hamstrings and by the correlation of changes in movement with EMG activity (Dimitrijevic, Faganel, Lehmkuhl, and Sherwood 1983).

A more sophisticated method to measure muscle tone is the isokinetic dynamometer (Cybex) (Knutsson 1985). This method has become useful for the measurement of muscle torque during passive stretch in spastic patients and to assess the effect of different treatment procedures (Knutsson 1983). By adding EMG recordings to measurements of muscle tone with the isokinetic dynamometer it is then possible to measure the torque as well as the amount of elicited motor unit activity (Fig. 12.3). Muscle tone, however, is only one of the features of upper motor dysfunctions in spinal cord injury patients and it is essential also to record activities of the corticospinal and bulbospinal tracts in addition to muscle hypertonia.

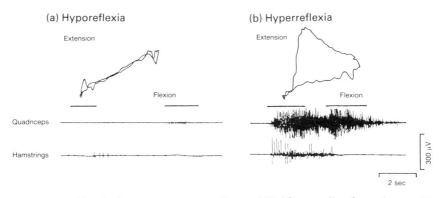

Fig. 12.3. Isokinetic dynamometer recording and EMG recording from the quadriceps and hamstrings during passive flexion and extension of the knee in a patient with hyporeflexia (a) and hyperreflexia (b).

Stimulation of the motor cortex

Merton and Morton (1980) demonstrated that it is possible to elicit muscle twitches of the forearm muscles by stimulating the motor strip over the scalp with a high-voltage sharp-rising stimulus (parameters for 5–800 V and 50 μsec). This technique of isolated motor cortex stimulation and induced muscle contraction of different muscle groups through the activation of fast-conducting fibres of the corticospinal tract is, in a way, very similar to that for the recording of electrical events above the sensory cortex after stimulation of peripheral sensory receptors or nerve fibres (Dimitrijevic, Prevec, and Sherwood 1983). It has been shown that the same technique can also be used for the stimulation of motor tracts at the spinal cord level (Swash and Snooks 1985). This new approach to examining the integrity of a portion of the long descending pathway has attracted the attention of many researchers, who have applied it to studies of the motor system in patients with neurological disorders. Recent technological developments resulting in the substitution of the electrical by the magnetic stimulator, which selectively depolarizes only the motor cortex and does not stimulate any peripheral sensory head structure, will definitely facilitate the use of this painless and non-invasive procedure for the examination of the conduction properties of the corticospinal tract in different neurological conditions (Barker and Jalinous 1985; Barker, Freeston, Jalinous, and Jarratt 1985).

Motor cortex stimulation in chronic post-traumatic spinal cord injury patients has not been widely used. We are now in the process of collecting data in spinal cord injury patients with anatomically and physiologically transected spinal cord, patients with morphologically and physiologically incomplete spinal cord, and patients with clinically complete but physiologically incomplete or so-called 'discomplete' spinal cord injury. By means of neurophysiological procedures which test segmental and suprasegmental functions in patients with clinically complete motor and sensory spinal cord injury it is possible to document, in the majority of patients, the influence of residual functions of bulbospinal pathways on segmental reflex activity. This finding has led us to propose a new term to describe this condition as 'discomplete' spinal cord injury.

We are using electrical stimulation of the motor cortex to elicit muscle twitches in the muscles of the upper and lower limbs by changing the anode position of the bipolar stimulating electrodes over the motor cortex (Fig. 12.4). The threshold for stimulation can be lowered by the volitional activation of responding muscle groups which, according to the report of Berardelli, Cowan, Day, Dick, and Rothwell (1985), is caused by the increased excitability of segmental mechanisms. This manœuvre can, therefore, diminish the discomfort of electrical stimulation.

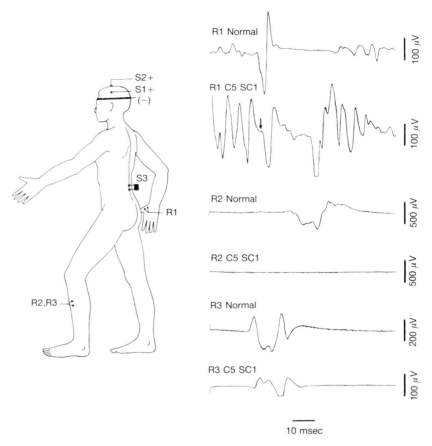

Fig. 12.4. Motor-evoked potentials in healthy and spinal cord injury subjects. The left of the drawing indicates the stimulation site of the motor cortex to elicit muscle twitch responses in the hand and dorsal flexors of the ankle (S1, S2) and the stimulation site of the lumbosacral portion of the spinal cord (S3). R1, R2, and R3 represent the appropriate recording sites. The EMG traces to the right show distal muscle EMG responses in a urologically healthy subject (first trace) and in a patient with spinal cord injury (second trace). When stimulated under S2 the R2 response is absent in spinal cord injury patients, whereas the R3 response is present in healthy subjects and spinal cord injury patients.

Long latency responses

The fast ascending and descending system of the central nervous system (CNS) can also be tested by using long latency responses, which are described as transcortical stretch reflexes and have recently been extensively studied and widely applied in different neurological conditions (Marsden, Rothwell, and Day 1983). In our experience, this long latency response is absent in patients with paralysis or with a very low degree of paresis, probably owing to the inability of the CNS to mediate a synchronized ascending input and descending output. Although these responses are important to test the slightest dysfunctions of the transcortical integrity within the long-loop stretch reflex, we are not yet certain since this investigation is still in progress. A limiting factor for studies of long-loop reflexes is also the presence of paralysis, which prevents us from eliciting a response since the long-loop reflex response requires the presence of volitional control which probably mediates the increase of excitability of segmental mechanisms. However, in contrast with the just described examination of the fast-conducting descending and ascending system, the testing of vibratory induced tonic reflexes in patients with chronic spinal cord injury is useful, even in severe degrees of paralysis, for it reveals the residual ability of the bulbar-reticular system to mediate the excitation necessary for input through the reticulospinal tract (Dimitrijevic, Spencer, Trontelj, and Dimitrijevic 1977) (Fig. 12.5).

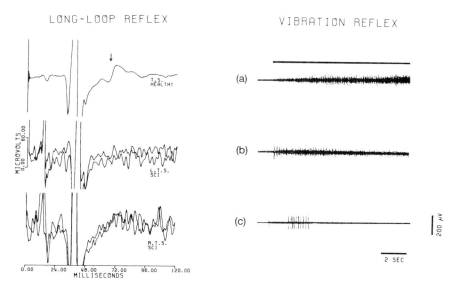

Fig. 12.5. Phasic and tonic long-loop reflexes. Three different traces are shown: (a) normal subject; (b) patient with moderate upper motor neuron dysfunction and functionally ambulatory with assistive devices; (c) non-ambulatory patient, ambulatory only with walker. In these three subjects, the long latency response and the vibration reflex were tested in the triceps *surae* muscle. It is clear that the long latency response disappears much earlier than the vibratory-induced reflex.

Testing the somatosensory system

Recording the early component of cortical somatosensory-evoked potentials after stimulation of the peripheral nerves of the upper and lower extremities is the most widespread electrophysiological technique to test spinal cord dysfunction. Somatosensory cortical-evoked responses are more often used for the assessment of acute spinal cord injury and early prognosis than for the functions of the posterior column in chronic spinal cord injuries (Maynard, Reynolds, Fountain, Wilmot, and Hamilton 1979). This technique should be used as a frequent follow-up to monitor changes within the early period of spinal cord injury, particularly when electrical events are present and follow-up studies can show changes in quantitative value (Perot and Vera 1982). Cortical somatosensory-evoked potentials have also been widely used to monitor spinal cord functions during spinal surgery (Schramm 1985).

In chronic spinal cord injury patients, cortical somatosensory-evoked events are important in testing the fast-conducting system of the posterior columns, but their absence does not exclude the patient's perception of altered sensation (Dimitrijevic, Beric, and Lindblom, unpub. obs.). They are also very important to test the functional integrity of the posterior columns independently of the patient's collaboration and correlation of clinical findings for impaired perception of touch, pain, cold and warm, vibration, and sense of position. By adding the quantification of sensation for touch, vibration, and cold and heat to neurological findings for sensory deficits, it becomes possible to outline the extent of the impairment of functions of different ascending tracts and to establish the anatomical and physiological basis for elicited and spontaneous sensation. When somatosensory cortical-evoked responses are added to recordings of cervical, thoracic, lumbar, pre-synaptic, and post-synaptic spinal cord events (Dimitrijevic, Lehmkuhl, Sedgwick, and Sherwood 1980), it is then possible to obtain information about the integrity of the peripheral input to the corresponding spinal cord regions below the lesion and in the absence of functions of the long ascending tracts (see Fig. 12.6).

Conclusion

When the spinal cord is integrated with the other parts of the nervous system in neurologically healthy subjects the motor, sensory, and autonomic functions are highly integrated and their interaction is harmonious and conducive to homeostasis. Spinal cord lesion results in structural deterioration and functional disintegration which cause different neurological malfunctions, dysfunctions, and deficits. However, these deficits are a consequence of the partial presence or absence of long ascending, descend-

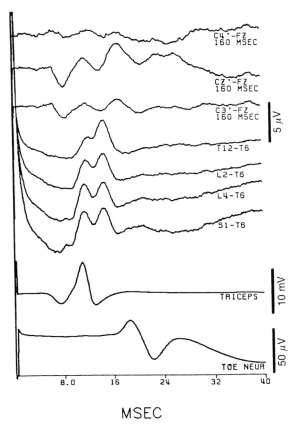

Fig. 12.6. The recording site of 128 averaged responses to tibial nerve stimulation at the *fossa poplitea*. The strength of the stimulus applied to the tibial nerve is monitored by the H-reflex and M-wave of the triceps *surae* and the electroneurogram of the big toe. Somatosensory-evoked responses are recorded over the lumbosacral portion of the spine and the somatosensory responses over the cortex.

ing, and propriospinal structures. Therefore, it is essential for the evaluation of motor and sensory dysfunctions in chronic spinal cord injury patients to define which residual structures have been definitely injured and which have not been affected and are still functioning. The clinical neurological evaluation can describe the neurological deficit, but clinical neurophysiology can elucidate which part of the ascending and descending system of the spinal cord still functions by using the previously described method to record polyelectromyographic activity during reflex, automatic, and volitional motor activity, to measure muscle tone, and to stimulate the cortex or spinal cord and the somatosensory-evoked potentials. Unfortunately, we do not yet have the possibility to monitor the activity of numerous distinct corti-cospinal, bulbospinal, spinobulbar, and spinocortical pathways. On the other hand, we might be able to record in a clinical environment several functions of ascending and descending fibres of the fast- or even slow-conducting system. These recordings should be used in routine clinical work along with the continuously developing techniques for the selective assess-ment of different spinal cord tracts, which will establish the neurobiological and neurophysiological basis for the restoration of lost functions in patients with chronic spinal cord injury.

By applying contemporary neurophysiological techniques in chronic spinal cord injury patients we have learned that, even in patients with neurological evidence of complete lesions, the functions of different spinal cord tracts are partially preserved, and that it is possible to modify these functions through physiological, pharmacological, or even surgical pro-cedures. This knowledge has led us to propose a number of restorative neurological procedures based on clinical neurophysiological assessments. Moreover, we are changing our attitude towards wheel-chair confined patients.

The two sculptures in Fig. 12.7 illustrate this point. The sculpture to the left is a neurologically healthy subject and the one to the right represents the patient's perception of paralysis. The body is different, it is square but

Fig. 12.7. Two sculptures. Neurologically healthy subject on the left, accompanied by 'square body' on the right, symbolizing the proposed different perceived image of wheelchair-'bound' spinal cord injury patient.

present and erect. Further work in tracing and documenting disintegrating spinal cord functions after injury might contribute in the near future to the assessment of spinal cord dysfunctions, not only from the standpoint of the neurologist and clinical examiner but also from that of spinal cord injury patients.

Acknowledgements

The author is grateful to the members of the Division of Restorative Neurology and Human Neurobiology, Baylor College of Medicine for their support and contribution, and wishes to express his deepest appreciation to the Vivian L. Smith Foundation for Restorative Neurology for its generous support.

References

Barker, A. T. and Jalinous, R. (1985). Non-invasive magnetic stimulation of human motor cortex. *Lancet* i, 1106–7.

Barker, A. T., Freeston, I. L., Jalinous, R., and Jarratt, J. A. (1985). Motor responses to non invasive brain stimulation in clinical practice. *EEG Clin. Neurophysiol.* **61**(3), S70.

Berardelli, A., Cowan, J. M. A., Day, B. L., Dick, J., and Rothwell, J. C. (1985). The site of facilitation of the response to cortical stimulation during voluntary contraction in man. *J. Physiol.* **360**, 52.

Dimitrijevic, M. R. (1984). Neurocontrol of chronic upper motor neurone syndromes. In *Electromyography in CNS disorders: central EMG* (ed. B. Shahani), pp. 111–27. Butterworth, Sevenoaks, Kent.

Dimitrijevic, M. R. and Faganel, J. (1985). Motor control of the spinal cord. In *Recent achievements in restorative neurology: upper motor neuron functions and dysfunctions. Vol. 1* (ed. J. Eccles and M. R. Dimitrijevic), pp. 152–62. Karger, Basel.

Dimitrijevic, M. R. and Sherwood, A. M. (1980). Spasticity: medical and surgical treatment. *Neurology* **30**(2), 19–27.

Dimitrijevic, M. R., Prevec, T. S., and Sherwood, A. M. (1983). Somatosensory perception and cortical evoked potentials in established paraplegia. *J. Neurol. Sci.* **60**, 253–65.

Dimitrijevic, M. R., Faganel, J., Lehmkuhl, L. D., and Sherwood, A. M. (1983). Motor control in man after partial or complete spinal cord injury. In *Motor control mechanisms in health and disease, advances in neurology. Vol. 39* (ed. J. E. Desmedt), pp. 915–26. Raven Press, New York.

Dimitrijevic, M. R., Lehmkuhl, L. D., Sedgwick, E. M., and Sherwood, A. M. (1980). Characteristics of spinal cord evoked responses in man. *Appl. Neurophysiol.* **44**, 119–25.

Dimitrijevic, M. R., Spencer, W. A., Trontelj, J. V., and Dimitrijevic, M. M. (1977). Reflex effects of vibration in patients with spinal cord lesion. *Neurology (Minneapolis)* **27**, 1078–86.

Knutsson, E. (1983). Analysis of gait and of isokinetic movements for evaluation of antispastic drugs or physical therapies. In *Motor control mechanisms in health and disease* (ed. J. E. Desmedt), pp. 1013–34. Raven Press, New York.

Knutsson, E. (1985). Assessment of motor function in spastic paresis and its dependence on paresis and different types of restraint. In *Recent achievements in restorative neurology: upper motor neuron functions and dysfunctions. Vol. 1* (ed. J. Eccles and M. R. Dimitrijevic), pp. 199–210. Karger, Basel.

Malezic, M., Dimitrijevic, M. M., Dimitrijevic, M. R., and Sherwood, A. M. (1984). Study of suprasegmental and segmental control of muscle innervation in ambulatory spinal cord injury patients. Paper presented at the Eighth International Symposium on ECHE, Dubrovnik, Yugoslavia.

Marsden, C. D., Rothwell, J. C., and Day, B. L. (1983). Long latency automatic responses to muscle stretch in man: origin and function. In *Motor control mechanisms in health and disease* (ed. J. E. Desmedt), pp. 509–39. Raven Press, New York.

Maynard, M. M., Reynolds, G. G., Fountain, S., Wilmot, C., and Hamilton, R. (1979). Neurological prognosis after traumatic quadriplegia. *J. Neurosurg.* **50**, 611–16.

Merton, P. A. and Morton, H. B. (1980). Stimulation of the cerebral cortex in the intact human subject. *Nature* **285**, 227.

Perot, P. L. and Vera, C. L. (1982). Scalp-recorded somatosensory evoked potentials to stimulation of nerves in the lower extremities and evaluation of patients with spinal cord trauma. In *Evoked potentials. Annals of the New York Academy of Sciences. Vol. 388* (ed. I. Bodis-Wollner), pp. 359–68. New York Academy of Sciences, New York.

Schramm, J. (1985). Spinal cord monitoring: current status and new developments. *Central Nerv. Syst. Trauma* **2**(3), 207–27.

Swash, M. and Snooks, S. (1985). Motor conduction velocity in the spinal cord. *J. Physiol.* **360**, 50.

The use of motor cortical stimulation to monitor spinal cord function during surgery

J. C. Rothwell

Introduction

High-voltage electrical stimulation over central areas of the scalp can produce short-latency activation of muscles on the contralateral side of the body. Responses are particularly large in the distal muscles of the arm and hand, and probably are mediated by rapidly conducting axons of cortico-motoneuronal cells of the cerebral motor cortex. The ability to activate central motor pathways by external stimuli can be used to monitor function in spinal motor pathways during surgery. Bipolar epidural recording electrodes can record descending waves of activity produced by a single scalp stimulus. This technique for monitoring central motor tracts may be a useful adjunct to conventional monitoring of sensory pathways during spinal surgery.

Cortical stimulation in normal subjects

In 1980, Merton and Morton described a technique which allowed the brain of conscious subjects to be stimulated through the skull with high-voltage electrical shocks. The technique appears to be successful because the high voltage decreases the resistance of intervening tissues between the electrode and the brain so that current can penetrate the scalp more easily (Hill, McDonnell, and Merton 1980). Stimulation is produced by a capacitative discharge of about 500 V with a time constant of 50 or 100 μs. Peak currents reach values of about 500 mA. Such stimuli are completely safe and have been used on many hundred, fully conscious, co-operative, normal volunteers and neurological patients without ill-effects. More recently, a new method of stimulation has been introduced (Barker, Jalinous, and Freeston 1985) which relies on a rapidly changing magnetic field to induce eddy currents in the brain. This magnetic neural stimulator has the advantage over electrical stimulation in that it is possible to activate deep structures without producing a massive direct twitch of the underlying scalp muscles.

As a result it is more comfortable than electrical stimulation (though this is scarcely a consideration with anaesthetized patients).

Using either method of stimulation it is possible to evoke twitches in contralateral muscles by stimulating over the motor cortex. Stimulation usually is given whilst the subject is performing a small background contraction since this decreases the threshold for an evoked muscle contraction. Figure 13.1 shows a typical example of responses evoked in the biceps and thenar muscles by electrical stimulation over the contralateral scalp.

The latency of the response comprises conduction delays in both central and peripheral pathways. Estimation of the delay within the central nervous system (CNS) can be made by subtraction of the time taken for the impulses to travel from spinal cord to muscle. This can be measured either by conventional F-wave techniques, or by using the high-voltage stimulator to activate motoneurons directly in the cervical or lumbar enlargements. Stimulation over the appropriate segment of cord is believed to activate motor axons at the exit zone or even within the ventral horn of the spinal cord (Mills and Murray 1986). As such it provides a simple method of obtaining peripheral conduction delays for all muscles of the arm or leg. Since the stimulus activates peripheral motor axons there is no need to

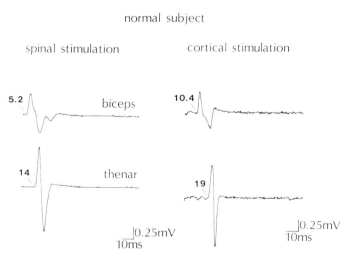

Fig. 13.1. Surface EMG responses from the biceps *brachii* and thenar muscles of a normal subject following high-voltage electrical stimulation of the cervical cord (left traces: T1 cathode, C5 anode) and contralateral scalp (right traces: vertex cathode, anode 7 cm lateral). Responses are averages of three trials each. Cortical stimulation given whilst the subject was exerting a small (5 per cent) background voluntary contraction; spinal stimulation given at rest. Latencies to onset of action potentials are shown in ms.

contract the muscle under test. Maximal responses are easily elicited in relaxed muscles. Figure 13.1 illustrates a typical example. In this subject, subtraction of peripheral conduction times gave a value of about 5 ms for the central conduction time from brain to low cervical cord.

Several lines of evidence suggest that the scalp stimulus activates large-diameter, rapidly conducting, corticomotoneuronal cells at the cortical or just subcortical level: (1) the central conduction delays are extremely short; (2) there is preferential activation of the hand and distal forearm muscles; (3) there is a rough somatotopy of stimulation sites on the scalp (see Rossini, Marciani, Caramia, Roma, and Zarola 1985), indicating that the stimulus activates superficial rather than deep structures.

Cortical stimulation as a monitoring technique

All operations on the spinal column carry with them some risk of damaging the cord. Present techniques of monitoring cord function all rely on recording activity in sensory pathways. Usually a peripheral nerve in the leg is stimulated and the ascending sensory potentials recorded from various places in its pathway through the CNS. These include the surface of the spinal cord, intervertebral ligaments, exposed spinous processes, or even from the scalp. Although these methods may monitor function successfully in sensory tracts, there is some evidence that it may in addition be important to monitor function in motor pathways. The difference in blood supply to the anterior and posterior parts of the spinal column raises the possibility of producing isolated ischaemic damage to either sensory or motor tracts (Turnbull, Brieg, and Hassler 1966). Indeed, there are a number of clinical reports showing the failure of sensory monitoring to reveal motor deficits produced during operations (Raudzens 1982).

Despite these findings, only Levy and York (1983) have attempted to assess function in motor tracts. Initially they did not have access to the high-voltage stimulator of Merton and Morton (1980) and used two different methods to activate motor tracts. These were direct stimulation of the dorsolateral part of the spinal column which had been exposed during surgery (Levy and York 1983), and transcranial (palate-to-scalp) stimulation of brain-stem pathways (Levy, York, McCaffrey, and Tannzer 1984). Both techniques produced descending activity in the spinal cord, but there was no guarantee that the signals were not contaminated by antidromic activity in sensory pathways.

The new electrical and magnetic cortical stimulators probably stimulate at or near the cortical level. Thus there can be no possibility of evoking antidromic descending activity in sensory tracts of the spinal cord. The technique is illustrated in Fig. 13.2. Scalp stimulation is used to evoke a descending volley in spinal motor pathways. This is recorded from bipolar

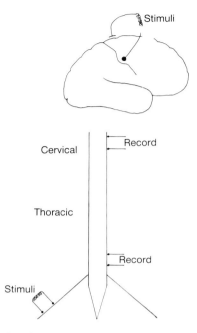

Fig. 13.2. Diagram showing the method of recording motor and sensory volleys from the spinal epidural space. Two electrodes are inserted at high and low levels of the cord, and stimuli are given either to the peripheral nerves in the leg, or to the cerebral cortex via electrodes on the scalp. Responses to single stimuli usually can be recorded following scalp stimuli, whereas it is necessary to average the responses to many hundred peripheral nerve stimuli.

epidural electrodes inserted into the spinal epidural space at different levels of the cord (for details, see Boyd, Rothwell, Cowan, Webb, Morley, Asselman, and Marsden 1986). The size and latency of the potentials can then be measured throughout the course of surgery to monitor the integrity of spinal motor tracts. As indicated in the figure, this may be combined with sensory monitoring by using the same electrodes to record ascending sensory volleys produced by stimulation of peripheral nerves in the leg. For successful recording of motor function, the patients must be fully paralysed so that there is no danger that the recordings are contaminated by muscle activity evoked in paraspinal muscles.

This technique has been used successfully to monitor spinal cord motor tracts in a series of children undergoing corrective surgery on the spinal column for scoliosis (see Boyd *et al.* 1986). Using epidural electrodes, it is relatively easy to record electrical responses at all levels of the spinal cord following stimuli given over the central areas of scalp. Typical results from a

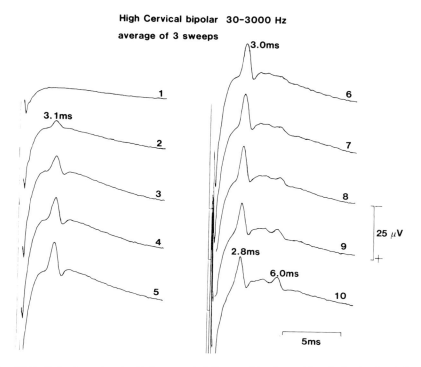

High Cervical bipolar   30-3000 Hz
average of 3 sweeps

3.0ms

3.1ms

1
2
3
4
5

6
7
8
9
10

2.8ms

6.0ms

25 μV

5ms

Fig. 13.3. Spinal cord potentials recorded from a bipolar epidural electrode at the level of the mid-cervical cord following scalp stimulation at different intensities. Intensities are given in arbitrary, but linear, units from 1 to 10. Maximum intensity (10) gives a 500 V capacitative discharge with a time constant of 50 μs. A short latency negative wave appears at low stimulation intensities, which increases in size and decreases in latency as the stimulation strength increases. The amplitude of the wave saturates at intensities 6–10, and a smaller later negative wave appears with a peak latency of about 6.0 ms. Traces are the average of three sweeps, recorded with filter settings shown. Stimulation occurs at the start of each record (see artefact).

cervical cord electrode are shown in Fig. 13.3. At low intensities of stimulation, a short-latency potential appears. This increases in size and decreases in latency as the intensity is increased. At high intensities, later waves appear with about twice the latency of the early potentials. We think that these responses represent action potentials descending in spinal motor tracts, because (1) the responses have a relatively short duration when recorded with open-filter settings, suggesting that they are not synaptic potentials, and (2) movement of the electrode in the epidural space changes the latency of the potential. This is illustrated in Fig. 13.4 (left). When the electrode was moved 2 cm distal, the latency of the potential increased by

Bipolar Epidural 30-3000 Hz
average of 5 sweeps

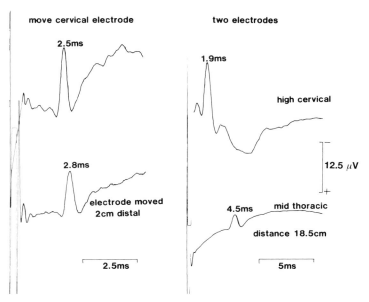

Fig. 13.4. Left, effect of moving a bipolar recording electrode in the cervical epidural space on the latency of the first negative potential produced by scalp stimulation. A latency difference of 0.3 ms over 2 cm gives a conduction velocity of 67 m/s. Traces are the average of five sweeps. In this illustration, scalp stimulation intensity was submaximal. Right, descending motor volleys recorded from two bipolar epidural electrodes in the high cervical and mid-thoracic spaces of a single patient. The interelectrode distance was 18.5 cm as measured over the vertebral bodies, and the conduction delay was 2.6 ms, giving a conduction velocity of 71 m/s. The potentials are the average of five sweeps. (From Boyd *et al.* 1986, with permission.)

0.3 ms. Assuming a travelling wave of activity, this gives a descending conduction velocity of 67 m/s.

In most patients the responses are relatively large, especially at cervical level. This is presumably because the stimulation site is close, and there is relatively little temporal dispersion of the volley. Because of this it is not necessary to record very many potentials to obtain a reproducible average response. This is in contrast to recording the sensory ascending volley, which usually requires averaging of 512 or more sweeps.

In different subjects, and at different levels of the cervical cord, the latency of the initial volley varied from 1.7 to 3.8 ms. As indicated above, measurements on normal subjects gave values of about 5 ms for the central conduction time from scalp stimulation to low cervical cord. However, this value includes: (1) delays due to slowing of conduction in the terminal sections of corticospinal axons; (2) delay at the corticomotoneuronal synapse; (3) the time taken to initiate an action potential in the spinal motoneurons and to conduct this to the point at which spinal stimulation occurs. In addition, it should be remembered that all the intraoperative data was obtained on children under the age of sixteen years, whose central conduction times may be slightly shorter than those in adults. Taking these factors into account there is good agreement between the two latencies, and this suggests that the descending volley recorded from the surface of the cord during surgery is the same as that responsible for activating spinal motoneurons in intact man.

Two other features of the evoked spinal activity deserve mention. First, there is a decrease in latency of the initial volley at high intensities of stimulation. This could be due either to recruitment of faster conducting axons or to spread of the stimulation current to more distal parts of the cortical axon. If the decrease in latency were due to the latter possibility, then a decrease of 0.3 ms (as in Fig. 13.3) in a fibre conducting at 60 m/s would imply that the current could spread 18 mm into the depths of the cortical white matter at maximum intensity. The second point which deserves discussion is the presence of small later waves in the epidural records. These may have been due to (i) activity in indirect pathways, (ii) activity in slower conducting axons of the corticospinal tract, or (iii) repetitive activity in the same large-diameter corticospinal neurons which were responsible for the initial descending volley. All three possibilities may contribute. However, the latter explanation is of particular interest since the early and late waves bear a very close resemblance to the D- and I-waves recorded from the pyramidal tract after anodal stimulation of the exposed cortex in primates. In these animals, D-waves have the shortest latency and lowest threshold, and are due to direct stimulation of large-diameter pyramidal tract axons. The later I-waves have a higher threshold and are caused by repetitive activity in the same pyramidal tract axons (Kernell and Wu Chien-Ping 1967).

To monitor the motor pathways from the whole length of cord it is necessary to record the descending volleys at low thoracic levels. Figure 13.4 (right) shows responses recorded simultaneously from both high cervical and mid-thoracic levels of the cord in one patient. At low levels, the responses are smaller, probably due to the absence of the large corticospinal projections to the arms. In addition, it is not possible to see any later waves of activity with certainty. If the distance between the recording electrodes is measured, such data can be used to calculate the spinal motor conduction velocity in man. Results obtained by this method, and by the method of moving a single recording electrode within the epidural space (see above),

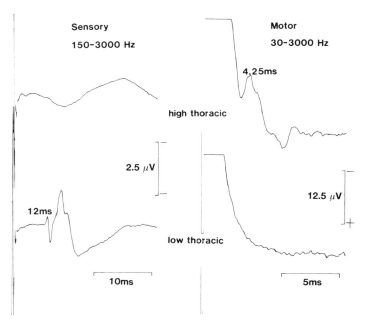

Fig. 13.5. Ascending sensory volleys (left) and descending motor volleys (right) recorded from bipolar epidural electrodes in the high and low thoracic epidural space in a patient who had been rendered paraplegic at the T6 level in a previous operation. The volleys are not conducted past the region of the block. Sensory potentials are the average of 500 sweeps, motor potentials are the average of five sweeps. The sensory potentials have an abnormal morphology. (From Boyd *et al.*, 1986, with permission.)

give similar values of between 50 and 74 m/s in different individuals. These values compare closely with those obtained for corticospinal tract conduction velocity in primates (see Phillips and Porter 1977).

Finally, an opportunity was taken to record from the epidural space of a patient who had been rendered paraplegic at the T6 level after a previous operation. Figure 13.5 shows both the ascending sensory potentials and the descending motor potentials recorded from two electrodes inserted above and below the site of damage. Sensory volleys were seen only at the lower electrode. Motor volleys were seen only at the upper electrode. Thus the technique is capable of demonstrating lesions in spinal motor pathways.

Summary

The technique of electrical (and the newer magnetic) brain stimulation evokes descending waves of activity in the spinal cord of anaesthetized patients. These volleys can be recorded from electrodes inserted into the spinal epidural space, and provide a means of monitoring spinal motor function during the course of spinal surgery.

Acknowledgements

This work was supported by the Medical Research Council. The majority of data reported here was collected in association with Dr S. G. Boyd of the Children's Hospital, Great Ormond Street, London.

References

Barker, A. T., Jalinous, R., and Freeston, I. L. (1985). Non-invasive magnetic stimulation of human motor cortex. *Lancet* i, 1106–7.

Boyd, S. G., Rothwell, J. C., Cowan, J. M. A., Webb, P. J., Morley, T., Asselman, P., and Marsden, C. D. (1986). A method of monitoring function in spinal motor pathways during scoliosis surgery with a note on motor conduction velocities. *J. Neurol. Neurosurg. Psychiat.* **49**, 251–7.

Hill, D. K., McDonnell, M. J., and Merton, P. A. (1980). Direct stimulation of the adductor pollicis in man. *J. Physiol.* **305**, 9–10.

Kernell, D. and Wu Chien-Ping (1967). Responses of the pyramidal tract to stimulation of the baboon's motor cortex. *J. Physiol.* **191**, 653–72.

Levy, W. J. and York, D. H. (1983). Evoked potentials from the motor tracts in humans. *Neurosurgery* **12**, 422–9.

Levy, W. J., York, D. H., McCaffrey, M., and Tannzer, F. (1984). Motor evoked potentials from transcranial stimulation of the motor cortex in humans. *Neurosurgery* **15**, 287–302.

Merton, P. A. and Morton, H. B. (1980). Stimulation of the cerebral cortex in the intact human subject. *Nature* **285**, 227.

Mills, K. R. and Murray, N. M. F. (1986). Electrical stimulation over the human vertebral column: which neural elements are excited? *Electroencephal. Clin. Neurophysiol.* **63**, 582–9.

Phillips, C. G. and Porter, R. R. (1977). *Corticospinal neurones*. Academic Press, London.

Raudzens, P. (1982). Intraoperative monitoring of evoked potentials. *Ann. N.Y. Acad. Sci.* **388**, 308–26.

Rossini, P. M., Marciani, M. G., Caramia, M., Roma, V., and Zarola, F. (1985). Nervous propagation along 'central' motor pathways in intact man: characteristics of motor responses to 'bifocal' and 'unifocal' spine and scalp non-invasive stimulation. *Electroencephal. Clin. Neurophysiol.* **61**, 272–86.

Turnbull, I., Brieg, A., and Hassler, O. (1966). Blood supply of cervical spinal cord in man: a microangiographic study. *J. Neurosurg.* **24**, 951–65.

Index